# MAMMOTHS

## Giants of the Ice Age

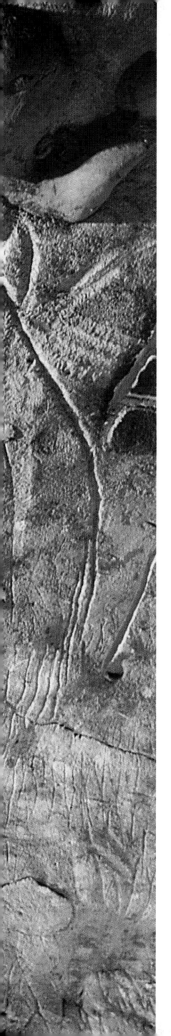

# MAMMOTHS
## Giants of the Ice Age

Adrian Lister
and
Paul Bahn

Revised Edition
Foreword by Jean M. Auel

中

University of California Press

Berkeley   Los Angeles

In memory of Barbara and Stanley Lister
and of Guy Bahn

University of California Press, one of the most distinguished university presses in the United States, enriches lives around the world by advancing scholarship in the humanities, social sciences, and natural sciences. Its activities are supported by the UC Press Foundation and by philanthropic contributions from individuals and institutions. For more information, visit www.ucpress.edu.

University of California Press
Berkeley and Los Angeles, California

Published by arrangement with Marshall Editions
The Old Brewery, 6 Blundell Street, London, N7

Cataloging-in-Publication Data for this title is on file with the Library of Congress

ISBN 978-0-520-25319-3 (cloth: alk. paper)

Manufactured in Singapore by Star Standard Industries (Pte) Ltd.

17 16 15 14 13 12 11 10 09 08
10 9 8 7 6 5 4 3 2 1

Page 1: *One of the most important mammoth finds in recent years is this well-preserved head, excavated in Yakutia, Arctic Siberia.*

Pages 2–3: *A dramatic engraving of a woolly mammoth from the cave of Rouffignac in France not only captures the distinctive physical shape of the animal but also suggests something of its imposing character. Its real size is 2ft 7in (80cm) high by 4ft (1.2m) long. The lumps protruding from the rock face are nodules of flint.*

*Above: A series of posters was issued in the last century in Russia offering rewards for reports of mammoth carcasses in Siberia.*

# CONTENTS

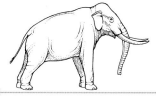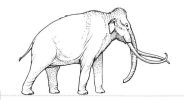

# FOREWORD

S ince few of us will ever have the opportunity to go to Africa or India to see elephants in their natural environment, it is only at zoos or circuses that we can gaze in wonder at these fantastic creatures—the largest to walk the Earth. But at one time, early relatives of elephants roamed the northern continents and grazed in places that are now the busy streets of major cities. Imagine how wonderful it would be to see those huge exotic animals, clothed in fur and sporting magnificent tusks.

In *Mammoths* Adrian Lister and Paul Bahn have given us the chance, or at least the next best thing to it. How I wish this readable book had been available when I was trying to find information about woolly mammoths for my books of Ice Age fiction, instead of having to search through dozens of highly technical books and papers.

Currently, in the popular perception—and in toy stores—mammoths are lumped together with dinosaurs as extinct animals. As the reader of this book will discover, dinosaurs died off 65 million years ago, and mammoths arose in tandem with humans around four or five million years ago. With all the attention that is currently being paid to extinct dinosaurs, it's time these equally intriguing behemoths got their share. *Mammoths* has set the record straight.

Professionals, science teachers, curious adults, and, though it is not specifically written for them, even children—anyone who is interested in learning the details of how these fascinating animals fit into the overall picture—will welcome this book. Within its pages, readers will discover how the European woolly mammoths were uniquely adapted to their cold world, and the differences between them and the American variety. They will learn that mammoths were cousins of the African and Asian elephants, not their ancestors. They will find out how we have come to know so much about mammoths, including their behavior.

*A carving of a mammoth* made of mammoth bone–probably a patella (kneecap) or a vertebra–was discovered at Avdeevo, in Russia. This site dates to between 22,000 and 12,000 years ago.

In this comprehensive and beautifully illustrated volume, each of the frozen specimens found in the ice and permafrost of the north is described, dispelling a few myths along the way. For example, except for dogs and other carnivores, no one is known to have actually eaten frozen mammoth meat, although it is deep red in color like beef. Scientists have extracted their DNA and are using it, along with careful study of their physical characteristics, to tell whether mammoths were more closely related to Asian elephants or the African variety.

From the paintings and engravings made by our own Ice Age ancestors, we perceive how they saw these massive and inspiring animals that shared their frozen world, and how important mammoths were to more than their survival. They lived in dwellings constructed of mammoth bones, and burned bones for fuel. They made tools from cleavers to needles to boomerangs out of mammoth tusks, and created art from them as well. But their art was more than carved objects. Early modern humans created music from mammoth bones, and probably used them in joyous celebration and ceremony.

In weaving together all the various strands of this fascinating story, Lister and Bahn have performed a great service for all of us. We have learned about more than the life of the mysterious mammoth, we have learned something about ourselves as well.

*A horse made of mammoth ivory, found at Vogelherd in Germany, was carved some 30,000 years ago, and is one of the oldest sculptures in the world.*

Jean M. Auel
Author of *The Mammoth Hunters* and other best-selling novels about the Ice Age

# INTRODUCTION

*"Round a bend in the path, the towering skull appeared, and we stood at the grave of the diluvial monster. The body and limbs still stuck partially in the masses of earth with which the corpse had been precipitated in a big fall from the bank of ice. We stood speechless in front of this evidence of the prehistoric world, which had been preserved almost intact in its grave of ice throughout the ages. For long we could not tear ourselves away from the primeval creature, the mere sight of which fills the simple children of the woods and tundra with superstitious dread."*

***A carving of a mammoth** in mammoth ivory was found at Předmostí in the Czech Republic. It is probably about 26,000 years old.*

These words of Eugen Pfizenmayer, a zoologist from the Russian Imperial Academy of Sciences, describing his dramatic encounter of 1901 on an expedition within the Arctic Circle, capture the special excitement and awe of finding a perfectly preserved frozen mammoth. We have both long been fascinated by mammoths, and through our research have been lucky enough to experience the thrills of excavating mammoth remains which had lain buried for thousands of years, and of seeing with our own eyes the depictions of these extinct animals left by our distant ancestors. It is this excitement and these experiences that we wish to share with readers through the pages of this book.

Among prehistoric creatures there is, we believe, no other species about which so many fascinating facts are known. In *Mammoths* we have attempted to cover all aspects of the mammoth—its life, death, and preservation, and its interactions with prehistoric people. In so doing, we have drawn on the latest discoveries and research, including our own work and that of many friends and colleagues. Our text is complemented by specially created illustrations that dramatically set the mammoths in their Ice Age habitats.

The book presents the mammoth story in a logical sequence, and begins by placing the animals in their evolutionary context, looking back at their origins and their links with similar creatures, notably the elephants, their nearest living relatives. This opening chapter also traces the migration of mammoths across the globe—a story that started in Africa some five million years ago and ended in places as far apart as Great Britain, Mexico, and Japan. We also explore the paradoxical evolution of "dwarf mammoths" on islands off California and elsewhere. Finally, we explain the latest DNA research that has helped us understand the mammoth's origins, and promises to provide further clues to the animals' appearance, although it is unlikely, at least in the foreseeable future, to produce a living mammoth.

Chapter two tells how mammoth remains are preserved, discovered, and unearthed. Here

*A perfectly preserved skull and tusks of the woolly mammoth came to light when a river bluff collapsed on the Bolshoy Khomus-Yurayakh River in northeast Siberia.*

readers can join Pfizenmayer and the "mammoth hunters" of old as well as the modern excavations that reveal new pieces of evidence that help to fill in the mammoth story. From single teeth in gravel pits to carcasses preserved entire and deep frozen in the permafrost, each has an important place in the record as a whole.

In chapter three we show how the information from mammoth remains is used to reveal everything we know about the mammoth as a living animal—its appearance and adaptations, its behavior and social life. Teeth and bones, skin and hair of mammoths—male and female, young and old—are used to help create pictures of the way these prehistoric creatures went about their daily lives.

Unlike the dinosaurs, mammoths coexisted with our early human ancestors, and chapter four is devoted to this relationship, beginning with the remarkable depictions of mammoths by the people of the Ice Age. Following this we reveal the many different ways in which humans used the animals' bones and tusks—for home building, for tools and ornament, and for ritual burial.

The final chapter investigates whether and how people may have hunted mammoths, questions how far this predation might have caused the extinction of the species, and looks at the role of climate in the mammoths' ultimate demise. Here

we describe the last living mammoths known to have survived on Earth—the miniature mammoths of Wrangel Island, which were probably still alive some 4,000 years ago, when Egyptian civilization was already established.

This edition has been completely revised from the earlier editions of 1994 and 2000. Numerous new discoveries of mammoth fossils have been made around the world, adding to our understanding of the animal and its environment. The latest research methods have been applied to the remains: state-of-the-art medical scanning technology allowing a peek at hidden structures; chemical and microscopic analysis of tusks revealing an animal's life history; DNA studies of the animals' genetics; and new methods of dating fossils and reconstructing their environment. We also illustrate many wonderful, newly-discovered cave drawings and objects, and summarize the extensive new research on the causes of the mammoth's extinction.

If we cannot entertain the hope of seeing mammoths alive, we can go a long way toward breathing life into their flesh and bones. This has been made possible by the efforts and discoveries of archeologists and paleontologists, together with the beautiful images left to us by the many prehistoric artists who saw these extraordinary animals in the flesh.

# ORIGINS

The ancestry of the mammoths can be traced back through time to about fifty-five million years ago. The last dinosaurs had become extinct about ten million years before the emergence of a group of mammals we know as the proboscideans, which soon evolved protruding tusks and extended, trunklike noses. Some fifty million years later, the first mammoths arose from this ancestral line.

Although they were related to modern elephants, mammoths were not their ancestors but came from a separate branch of the family tree. The first mammoths walked and browsed in the tropical woodlands of Africa, but later migrated into Europe and Siberia and eventually reached North America. With the increasing cold of the Ice Age, mammoths in northern latitudes became transformed, through evolution, into the familiar woolly mammoths, which were highly specialized for survival in Arctic temperatures. In North America, by contrast, a rather different species, the Columbian mammoth, evolved. And the isolated environments of islands also led to their own peculiar transformations: dwarf mammoths, some of which seem to have been the very last mammoths on Earth.

*Living relatives of the mammoth, today's elephants are the last survivors of a formerly widespread and diverse group of animals spanning 55 million years of evolution.*

# EUROPE'S FIRST MAMMOTHS

A group of male ancestral mammoths browses in a lush, forested area of southern Europe about 1.5 million years ago. This species, *Mammuthus meridionalis*, was the direct descendant of tropical ancestors which had migrated north out of Africa two million years previously.

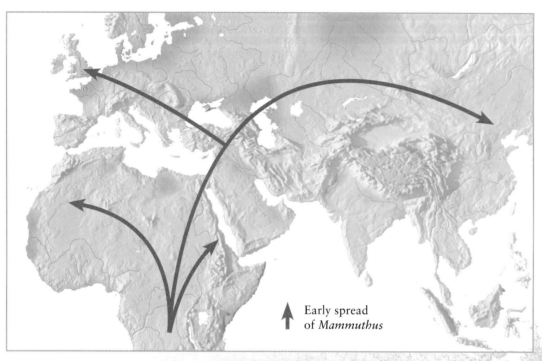

↑ Early spread of *Mammuthus*

**Europe in the Early Pleistocene** *enjoyed a mild climate for much of the time, and the forests included many species of animals and plants whose closest relatives now live farther south. One example is the porcupine, which today is found in North Africa and the Middle East.*

**At first sight, these ancestral mammoths** *look more like the elephants of today. Not yet adapted to the cold of the Ice Age, they lack the hairy coat of their descendants. However, several features mark them out as mammoths, in particular the single-domed head and the beginnings of a spiral twist to their tusks. Their ears were probably similar in size to those of modern Asian elephants.*

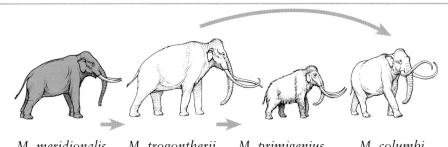

M. meridionalis    M. trogontherii    M. primigenius    M. columbi

**Mammuthus meridionalis**
*was the ancestral mammoth
from which all later
mammoths evolved:*
M. trogontherii *(the steppe
mammoth)*, M. primigenius
*(the woolly mammoth)*, and
M. columbi *(the Columbian
mammoth)*.

**Unlike their ultimate descendants,** *which were primarily
grass-eaters, ancestral mammoths were adapted to feeding
mostly on trees and shrubs. They ate the fruit and bark as
well as the leaves. Like modern elephants, these animals
almost certainly had a direct impact on their environment
through their habit of stripping the bark from trunks and
even knocking whole trees over.*

# MAMMOTHS OF THE STEPPES

A group of steppe mammoths, *Mammuthus trogontherii*, migrates along a river valley in central Europe around 600,000 years ago. The steppe mammoth was the largest of all the mammoths, big males reaching a height of 14 ft (4.3 m) at the shoulder and weighing at least 10 tons.

*The steppe mammoths' habitat* included large areas of grassy vegetation like the modern steppes—hence the species' name. Grass provided the bulk of its fodder. However, trees and shrubs were to be found along river valleys and in other sheltered areas, and their leaves and branches supplemented the mammoths' diet.

*Steppe mammoths probably showed* the first signs of the development of a thick coat, the reduction in size of the ears and tail, and other features later seen in the woolly mammoth. Given the habits of modern elephants, it is fair to assume that mammoth groups came together to migrate once or several times each year to search for new feeding grounds.

ORIGINS

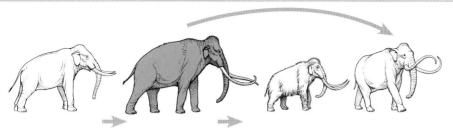

M. meridionalis   M. trogontherii   M. primigenius   M. columbi

**M. trogontherii** *(the steppe mammoth) was an evolutionary link between the earlier M. meridionalis (the ancestral mammoth), from which it arose, and the later M. primigenius (the woolly mammoth), which was its descendant.*

*Many other large animals thrived in this steppe habitat. One was the broad-antlered moose. Some 7 ft (2 m) high, it was the largest deer ever to evolve. Its huge antlers spread out sideways from its head.*

*Mammoths were not the only species of elephant in the European Pleistocene. The straight-tusked elephant Palaeoloxodon antiquus (left), probably related to the Asian elephant Elephas, equalled the steppe mammoth in size. It generally lived in warmer, more forested areas than the mammoth, but the two sometimes coincided where there was both grassland and trees.*

# ADAPTED FOR THE COLD

With its hairy coat, sloping back and huge tusks, the woolly mammoth, *Mammuthus primigenius* has become a symbol of the Ice Age. 50,000 years ago was the middle of the last Ice Age and the heyday of the woolly mammoth.

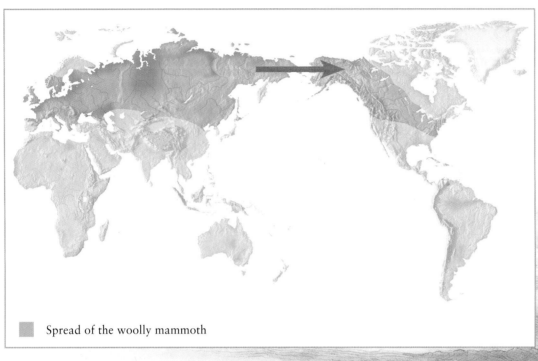

▮ Spread of the woolly mammoth

*The woolly mammoth probably arose in Siberia, but soon came to occupy a vast range, stretching from Ireland to the east coast of North America. Living south of the ice sheets, it inhabited a landscape of rich, grassy vegetation largely devoid of trees. Sharing this habitat were other now extinct species, such as the woolly rhinoceros and giant deer.*

*A light snowfall has obscured the vegetation, and one mammoth is using its tusks to clear the snow, revealing its grassy food. Contrary to the popular image of the mammoth habitat, snow was rarely heavy over much of its range.*

M. meridionalis    M. trogontherii    M. primigenius    M. columbi

**M. primigenius** *(the woolly mammoth) was descended from* M. trogontherii *(the steppe mammoth), which had in turn evolved from* M. meridionalis *(the ancestral mammoth).*

*At close quarters, the woolly mammoth, with its huge tusks, presented an imposing sight to any smaller animals, including humans. The details, such as its small ears, pointed trunk tip, and the length of its hair, are known from whole carcasses found in Siberia. The woolly mammoth represents the end point in a series of adaptations to the Ice Age habitat.*

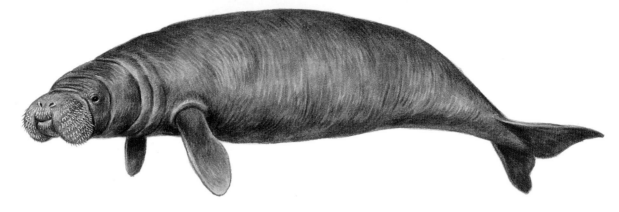

Mammoths rank with the dinosaurs as the most celebrated of prehistoric animals. The dinosaurs, however, were reptile-like animals and were extinct by 65 million years ago, while mammoths were mammals and did not appear until about 5 million years ago, lasting until only a few thousand years before the present. Mammoths are therefore very much closer to us in time and, indeed, coexisted with our human ancestors.

With the demise of the dinosaurs, the "Age of Mammals" began. Mammals had been in existence for a long time before that—their earliest remains date as far back as 200 million years—but during their early history they had remained as small, shrew-like animals living in the shadow of the dinosaurs. With the departure of the dinosaurs, mammals rapidly diversified into the many varied groups known today, as well as others that are now extinct. By about 50 million years ago, most of the major living mammalian groups had begun their evolution. These groups, known as orders, include the rodents, the carnivores, the whales, the primates, the bats, and so on.

The mammoth was a member of the mammalian Order Proboscidea. This name comes from the Greek *proboskis* and refers to the characteristic trunk. Today, only two or three species of proboscideans remain—the Asian elephants and the African (forest and savannah) elephants. However, the Proboscidea were previously much more diverse—remains of up to 160 different species have been found worldwide.

*Dugongs, which inhabit the coasts around Southeast Asia and the Indian Ocean, are among the closest living relatives of mammoths and today's elephants. They have small tusks formed from the incisor teeth, but these usually remain hidden beneath the lips.*

The story of the evolution of the Proboscidea has been worked out by the study of fossils, and also of the remaining living species. Among other mammals, it seems that the closest living relatives of the Proboscidea are the Order Sirenia. These are the sea cows, manatees, and dugongs: large, barrel-shaped, slow-moving aquatic mammals which feed on sea grasses around tropical coasts. At first sight, they seem unlikely relatives of elephants, but at the time of their common ancestry, at least 60 million years ago, both groups were smaller, land-living (perhaps amphibious) mammals, which had not yet acquired their specialized adaptations. Detailed anatomical observation reveals their relatedness: for example, sirenians and elephants share a peculiar division at the top of the heart, and in both groups the mammary glands are between their forelegs instead of farther back as in most mammals. Somewhat more distantly related to both proboscideans and sirenians are hyraxes—small, herbivorous creatures of tropical Africa. Similarities of the ear and leg bones have suggested distant relatedness to elephants, and in recent years this has been corroborated by molecular studies of proteins and genes.

*Tree-dwelling hyraxes from Africa may bear little resemblance to elephants, but they have a number of physical features in common which suggest a shared ancestry. These include the anatomy of their ear and leg bones.*

*Amebelodon*

## TUSKS AND TRUNKS

The earliest known proboscideans arose about 55 million years ago, in the area of the former Tethys Sea, roughly where the Mediterranean is today. Lacking a trunk, and with small, hippolike tusks, many of the early forms (such as the piglike *Moeritherium*) were amphibious. Some, among a group known as the barytheres, became very large and ponderous.

> ## "Many distant ancestors of the mammoths had four tusks"

The earliest proboscideans to show evidence of a trunk and tusks were the deinotheres, which arose about 40 million years ago, and all later proboscideans bear these distinguishing traits. The trunk is formed by a fusion of the nose and upper lip; where it joins the skull there is a large, broad, keyhole-shaped nasal opening. The existence of this feature allows us to deduce the presence of a trunk in fossil forms of proboscideans; the trunk itself never appears in ancient fossils since it contains no bones. The tusks are greatly enlarged incisors—originally biting teeth near the front of the mouth. In some species, such as *Deinotherium*, the lower incisors formed the tusks. In others, such as the mammoths and the living elephants, it is the upper incisors. Many fossil proboscideans had both upper and lower tusks—a total of four in all.

The heyday of the Proboscidea was in the period of time known as the Miocene—between about 24 and 5 million years ago. Many different species evolved, broadly known as mastodonts, but grouped into several distinct families. The mammutid family included the American mastodon *Mammut*, which survived until about 11,500 years ago. This species therefore came to coexist with the mammoths, although it was only a distant cousin whose family history had taken a separate course 25 million years previously. Other families included the "shovel-tuskers" such as *Amebelodon*, with enormously expanded, flattened lower tusks, as well as a pair of smaller upper ones.

One particularly significant group of Miocene Proboscidea was formed by the stegodons of Southeast Asia and Africa, because it was from this group that the true elephants are believed to have arisen. Stegodons had long, straight tusks, and their molar teeth (the back, grinding teeth) show the beginnings of the ridged form that characterizes those of elephants.

*Deinotherium*

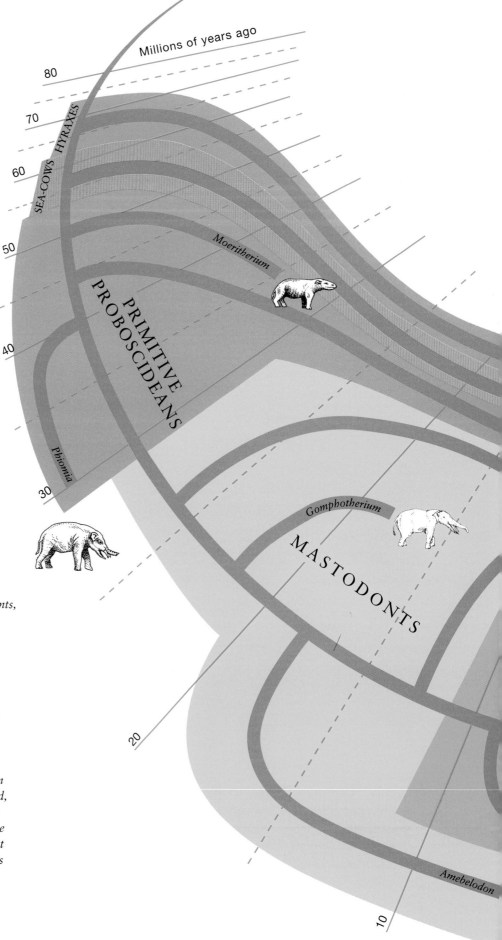

Millions of years ago

80

70

HYRAXES

SEA-COWS

60

50

*Moeritherium*

PRIMITIVE PROBOSCIDEANS

40

30

*Phiomia*

20

*Gomphotherium*

MASTODONTS

10

*Amebelodon*

**The ancestry of the mammoth** can
be shown in a simplified family tree.
The hyrax and dugong, closest living
relatives of the mammoth and elephants,
are nonetheless separated from them
by more than 50 million years of
proboscidean evolution, starting
with Moeritherium, and followed
by up to 160 fossil species, only a
few of which are shown here. Each
group's time of origin is approximate
and is estimated from its earliest
fossil dates together with its degree
of evolutionary advancement. For
example, Deinotherium is first known
from rocks about 25 million years old,
but its primitive features suggest that
its line diverged much earlier. The tree
reflects the latest genetic evidence that
the closest relative of the mammoth is
the Asian elephant.

# THE PROBOSCIDEAN FAMILY TREE

Tracing the Ancestry of the Mammoth

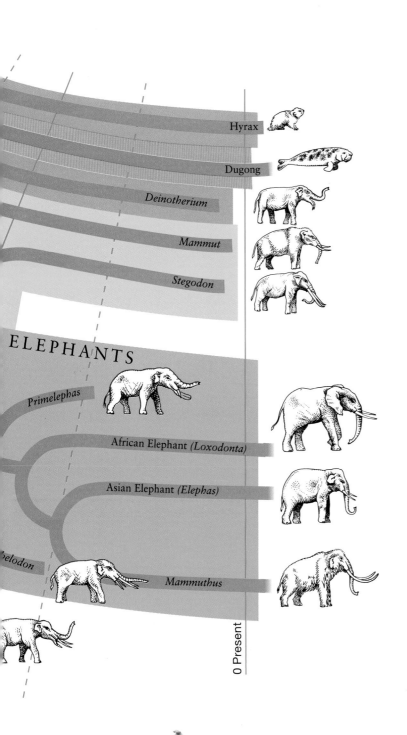

Hyrax

Dugong

*Deinotherium*

*Mammut*

*Stegodon*

ELEPHANTS

*Primelephas*

African Elephant *(Loxodonta)*

Asian Elephant *(Elephas)*

...elodon

*Mammuthus*

0 Present

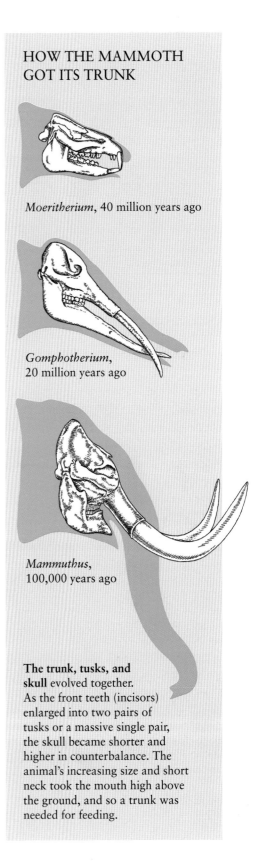

*Moeritherium*, 40 million years ago

*Gomphotherium*, 20 million years ago

*Mammuthus*, 100,000 years ago

**The trunk, tusks, and skull** evolved together. As the front teeth (incisors) enlarged into two pairs of tusks or a massive single pair, the skull became shorter and higher in counterbalance. The animal's increasing size and short neck took the mouth high above the ground, and so a trunk was needed for feeding.

## OUT OF AFRICA

The true elephants, including the two living species and the mammoths, form a family within the Proboscidea known as the Elephantidae. The crucible of elephant evolution appears to have been in Africa, for it is there that the earliest fossil representatives, and the evidence of the beginnings of diversification into different forms, are found. Only later did the elephants leave Africa to colonize other parts of the world.

Two distinctive features of mammoths and other elephants, in contrast to their predecessors, are the absence of enamel around the tusks and their ridged molar teeth (see pp.26–27). The tusks are composed of solid dentine, the hard tissue at the core of the incisor teeth. The earliest fossil forms with these features—and therefore the earliest true elephants—are the four-tusked *Stegotetrabelodon* and smaller *Primelephas*, which lived in Africa about 6 million years ago. Around the same time, still in Africa, the elephant line split into three main branches.

The first branch, *Loxodonta*, produced the living African elephant, *L. africana*. This lineage has remained in Africa since its origin, and has produced two forms, the larger savannah and smaller forest elephant, regarded as distinct species by some zoologists. The second branch, *Elephas*, eventually gave rise to today's Asian elephant, *E. maximus*. The fossil record shows that this line first diversified into a number of species in Africa before one of them migrated north and east into

## DIFFERENCES BETWEEN MAMMOTHS,

| SPECIES | **Woolly mammoth** *Mammuthus primigenius* | **American mastodon** *Mammut americanum* |
|---|---|---|
| HEIGHT | 9–11 ft (2.75–3.4 m) | 8–10 ft (2.4–3 m) |
| WEIGHT | 4–6 tons | 4–5 tons |
| BACK SHAPE | sloping | straight |
| FUR | dense | probably dense |
| HEAD | high single dome | low single dome |
| EAR | very small | unknown |
| TUSKS | curved and twisted | sometimes two pairs |
| TRUNK TIP | 1 short, 1 long "finger" | unknown |
| TAIL | short | medium |

*Elephants are the closest living relatives of the mammoth. Like today's elephants, the earliest mammoths evolved in a tropical climate— in contrast to the woolly mammoth, which acquired various adaptations to survive in an Arctic habitat.*

## MASTODONTS, AND ELEPHANTS

| African savannah elephant | Asian elephant |
| --- | --- |
| *Loxodonta africana* | *Elephas maximus* |
| 10–11 ft (3–3.4 m) | 8–10 ft (2.4–3 m) |
| 4–6 tons | 3–5 tons |
| saddle-shaped | humped |
| very sparse | sparse |
| low single dome | double dome |
| large | medium |
| gently curved | gently curved |
| 2 equal "fingers" | 1 "finger" |
| long | long |

Fossils of the earliest representatives of the mammoth line, which lived about 4–5 million years ago, were first identified in the 1920s. Named *Mammuthus subplanifrons*, they are found in South Africa, Ethiopia, and other countries in East Africa. Their remains are recognized as belonging to *Mammuthus* on the basis of features such as the spirally twisting tusks that are unique to mammoths. From the sediments in which they were found, and from plant fossils discovered with them, it is clear that they were living in a tropical environment, very different from that of their ultimate and most famous descendants, the woolly mammoths.

> ## "Mammoths and elephants roamed the Earth at the same time— they were close cousins"

*M. subplanifrons* is known to have survived until about 3 million years ago. Some time after this another mammoth species appeared in North Africa and was probably its descendant. Named *Mammuthus africanavus*, it seems to have been relatively small, with tusks that diverged more widely from the skull than is usual in other species of *Mammuthus*. This has led some paleontologists to suppose that it was an evolutionary "dead end," since the feature is not found in later species.

India and Southeast Asia, where it evolved into the Asian elephant still found there.

The third main branch of the elephant family was the line that was to lead to the mammoths (*Mammuthus*). For a long time it was unclear which living species of elephant was more closely related to the mammoth. Recent analysis of mammoth DNA (see pp.40–43) has resolved the issue in favor of the Asian elephant. This also means that although mammoths were close cousins of the living elephants, they were not their ancestors. For at least 4–5 million years, elephants and mammoths were contemporaries and were evolving separately. Only the relatively recent demise of the mammoths, a few thousand years ago, gives the false impression that mammoths were more ancient than, and perhaps therefore ancestral to, modern elephants.

*One of the world's oldest mammoth fossils, this molar of* Mammuthus subplanifrons *was found at Langebaanweg, South Africa, in deposits five million years old.*

# MAMMOTHS MOVE NORTH

Around 3 million years ago, the first mammoths appeared in Europe. Remains have been found at sites such as Montopoli, near Florence in Italy, in coastal deposits known as Red Crag in Suffolk, England, and most recently during excavations in the Russe region of northeast Bulgaria. Some scientists have suggested that they came from Africa via the western Mediterranean at the Strait of Gibraltar. However, the seaway already existed by then and a more probable route was via the Middle East and Turkey. These populations derived either from *Mammuthus africanavus* or, more likely, from direct descendants of *M. subplanifrons*.

The seemingly long distances become plausible when the timescale is considered. With an average spread of just 3 miles (5 km) per year, the 3,600 miles (6,000 km) from East Africa to northwestern Europe would be achieved in 1,200 years—a short period in the history of life on Earth. Migration into Britain was easy because the English Channel did not exist until half a million years ago.

The early European mammoth fossils, from between about 3.5–2.5 million years ago, are named *Mammuthus rumanus* because they were first identified from remains found in Romania. Important fossils of the species continue to be found in that country, but we know relatively little about this animal or its ecology.

Much better known is the next species in the line, *Mammuthus meridionalis*. Their fossils were first studied in 1825 by the Italian paleontologist Filippo Nesti, who did not recognize their affinity with mammoths and gave them the species name *Elephas meridionalis*.

Following the original discovery, the remains of *M. meridionalis* have been found in many European countries, including European Russia. Thousands of fossils have been collected around Florence over the past hundred years or so, and in the 1980s 15 individual skeletons were excavated at a lignite mine at Pietrafitta in central Italy. Isolated skeletons are known from other localities, such as Durfort in central France and Nogaisk in southern Russia, but, as with all fossil

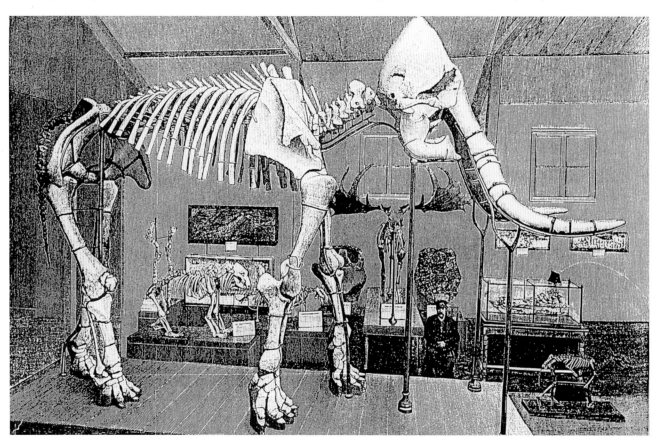

*A skeleton of an ancestral mammoth,* Mammuthus meridionalis, *was exhibited in the Paris Museum of Natural History in 1893. Excavated at Durfort in central France, it was the most complete specimen known at the time and formed the basis of early reconstructions of the ancestral mammoth.*

proboscideans, the most abundant finds are isolated teeth. For example, the British zoologist Sir Richard Owen wrote in 1846 that as many as 2,000 molars of *"Elephas" meridionalis* had been dredged from submarine deposits off the village of Happisburgh in Norfolk, England, in just 13 years.

From these finds a picture of *M. meridionalis* and its habitat emerges. Complete skeletons reveal an animal much bigger than a modern elephant—about 13 ft (4 m) high and probably weighing around 10 tons. The tusks were robust and showed the characteristic mammoth twist. Plant and other fossils found with the remains (see p.172) show that *M. meridionalis* was living in a time of mild climate, generally as warm as or slightly warmer than Europe experiences today. Deciduous mixed woodland provided its habitat and food, which comprised mostly tree-browse: oak, ash, beech, and other familiar European trees, as well as some that are now exotic to the region, such as hemlock, wing nut, and hickory. Further east, discoveries at Ubeidiyah (Israel) and Dmanisi (Georgia) show the early mammoth living in a partially open habitat with grassy areas, though subsisting on scattered trees and shrubs.

Judging from its climatic context, *M. meridionalis* probably lacked the dense fur of later mammoths and would have looked more like a typical elephant. However, the curved tusks and the rather pointed head marked it out as a mammoth. It lived alongside numerous other exotic animals such as porcupines, comb-antlered deer, zebralike horses, small rhinoceroses, and primitive pigs and cattle.

## THE STEPPE MAMMOTH

*Mammuthus meridionalis* survived until approximately 0.75 million years ago, but during its reign, starting about two million years ago, climatic changes were accelerating. These brought about shifts of adaptation in many mammalian species, including the mammoths. As global cooling intensified, much of the warm, forest habitat of the ancestral mammoth reverted to a more open, grassy landscape. It was almost certainly this change of conditions that led to the progressive evolution toward the familiar woolly mammoth, highly specialized for the Ice Age world.

Clear changes were first identified in European mammoth fossils dating to between 750,000 and 500,000 years ago, enough to identify a separate

*The mounted cast of a skelton of steppe mammoth* M. trogontherii *from Steinheim, Germany, on display in the Stuttgart Museum of Natural History.*

species. This was *Mammuthus trogontherii*, also known as the steppe mammoth because some of the best fossils, from Germany, were found associated with plant remains indicating a grass-dominated habitat.

Compared to those of the ancestral mammoth, the steppe mammoth's teeth had changed, reflecting the shift in diet: they have more enamel ridges to cope with tough, grassy food, and a higher crown to allow for greater wear through life.

Based on these fossils, it was assumed that *M. trogontherii* had evolved from *M. meridionalis* in Europe around 0.75 million years ago. Then, in 2001, mammoth remains from the Kolyma region of northeastern Siberia were found to be similar to those of *M. trogontherii*—but were a million or more years old. Subsequently, in 2005, excavations at the Maguanjou site in north-east China, which had produced evidence of the earliest humans in eastern Asia, yielded beautifully-preserved mammoth molars of *M. trogontherii* type. Dating of the locality produced an age of 1.66 million years, suggesting that the evolutionary change from ancestral to steppe mammoth began in eastern Asia, where there is evidence of cold winters and extensive grassy areas by that time, as well as forest.

# EVOLUTION OF MAMMOTH SKULLS AND TEETH

The 5-million-year transition from the African progenitor *Mammuthus subplanifrons*, to the forest-dwelling ancestral mammoth *M. meridionalis* and thence to the cold-adapted woolly mammoth *M. primigenius* involved a number of changes, such as the development of thick fur (see pp.84–85) and the reduction in the size of the ears and tail. One of the clearest pointers to this process of adaptation is the fossil teeth, always a good indicator of diet. Some 2.5 million to 1.5 million years ago, the ancestral mammoth had molars (see tooth evolution graph on opposite page) with comparatively low crowns and a small number of thick enamel ridges (usually no more than 12 to 14 on the back molar). This relatively modest chewing power was sufficient for a woodland diet, consisting mainly of the soft leaves of trees and shrubs.

Later fossils, which trace the transition to *M. trogontherii* and thence to *M. primigenius*, have progressively higher tooth crowns and more enamel ridges, around 18 to 20 in the back molars of *M. trogontherii*, and up to 26 in *M. primigenius*. This was linked to the shift to a grassy diet, which required more chewing power and wore down the teeth more rapidly—due to the minute silica particles in grass leaves and to the tendency of the animal to pick up grit when feeding. The American species *M. columbi* retained molars of similar form to *M. trogontherii*, from which it was probably derived.

This evolutionary sequence has been worked out by a comparison of teeth from successive time levels. Only accurately-dated fossils were used, so that they could be placed in their correct position in the sequence. It was also important that as many specimens as possible were examined from each level, so that normal variation between individual animals was not mistaken for evolutionary change. Finally, as far as possible, fossils were examined from across the geographical range of the species.

The average number of ridges in each sample of molar teeth has been plotted against the age of the samples to show the evolutionary trend. On the graph (opposite page), each species has been shown against its time of first appearance. The actual pattern, including geographical variation and movement, was more complicated (see main text and pp.30–31).

At the same time, there were changes in the skull and jaws. Partly to accommodate the higher-crowned teeth, the upper and lower jaws deepened. The top of the skull grew taller, providing a greater area of attachment for tendons and muscles running to the back. These acted to hold up the skull in counterbalance to the increasing weight of the tusks. The mechanical effort involved in moving such a weighty apparatus up and down was further assisted by the shortening of the skull from front to back, while the tusk sockets were now directed downward rather than forward. The result was the high but short skull which characterizes both the woolly and Columbian mammoths (top).

M. primigenius, *Siberia*

*In the evolution from ancestral mammoth (above) to woolly mammoth (below), the skull became higher and the face and jaw more compressed front to back.*

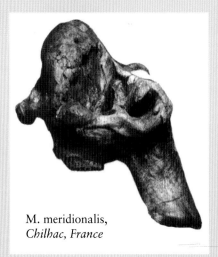

M. meridionalis, *Chilhac, France*

# MOLAR EVOLUTION

**primigenius:** *Balderton, UK*

**columbi:** *La Brea, California*

**trogontherii:**
*the earliest known
fossil of this species,
representing the biggest
single step towards the origin of
the woolly mammoth, discovered
at Maguanjou, China, in 2005*

Earliest appearance
(millions of years ago)

*primigenius* ✪

✪ *columbi*

*trogontherii* ✪

✪ *meridionalis*

✪ *rumanus*

✪ *subplanifrons*

Average number of enamel "plates" in molar

**meridionalis:** *North Sea*

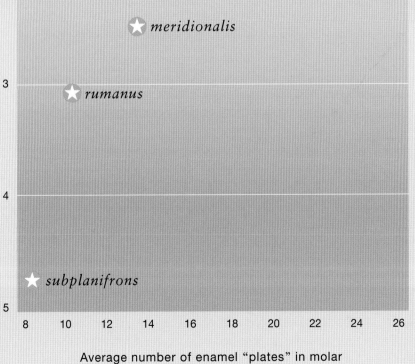

**rumanus:** *Cernatesti, Romania*

**subplanifrons:** *Langebaanweg, S.Africa*

*M. trogontherii* may, therefore, have originated in these northeastern regions of Eurasia about 2.0–1.5 million years ago and then, at about 750,000 years ago, moved into Europe as cool climates started to spread southward. It replaced its ancestor *M. meridionalis*, which became extinct. One or two fossil localities, intriguingly, appear to show the two species coexisting briefly, and rare fossils of intermediate form may suggest occasional hybridization (see p.31).

Complete skeletons of steppe mammoths are relatively rare. In Russia, two individuals, one male, one female, have been excavated from deposits about 600,000 years old near the northern shores of the Black Sea, the most recent in 1999. Of similar age, a very important find, on the Irtysh River of Western Siberia, was excavated in 1993. The skeleton of a male steppe mammoth, about 30 years old at death, was found together with remains of insects, plants, fishes, and rodents indicating a "forest steppe" habitat of extensive grassy areas with smaller areas of pine and other trees.

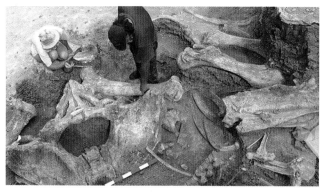

*The excavation of a steppe mammoth at Azov, the Black Sea.*

This was probably the most typical environment of the steppe mammoth. However, there is also evidence of the species living in warm, interglacial habitats with a more wooded or park-like vegetation of grasses and trees. An early example is the skeleton excavated in the 1990s at West Runton, Norfolk, UK (see p.75), around 700,000 years old. One of the latest comprises tusks and teeth found at Stanton Harcourt, Oxfordshire, UK (see p.71),

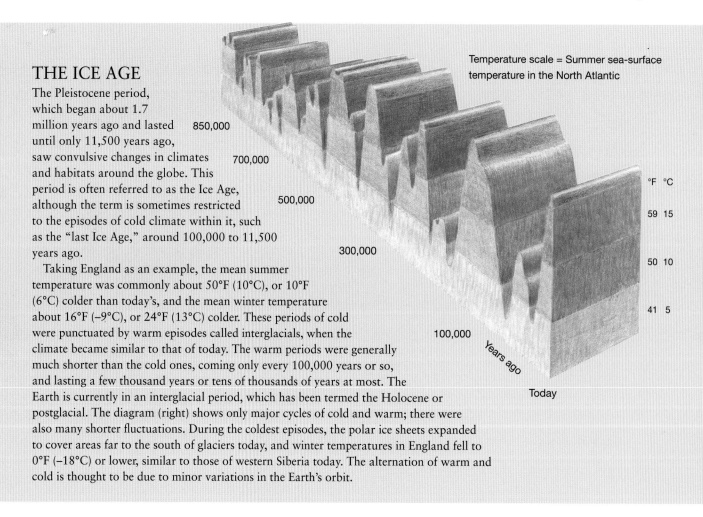

## THE ICE AGE

The Pleistocene period, which began about 1.7 million years ago and lasted until only 11,500 years ago, saw convulsive changes in climates and habitats around the globe. This period is often referred to as the Ice Age, although the term is sometimes restricted to the episodes of cold climate within it, such as the "last Ice Age," around 100,000 to 11,500 years ago.

Taking England as an example, the mean summer temperature was commonly about 50°F (10°C), or 10°F (6°C) colder than today's, and the mean winter temperature about 16°F (–9°C), or 24°F (13°C) colder. These periods of cold were punctuated by warm episodes called interglacials, when the climate became similar to that of today. The warm periods were generally much shorter than the cold ones, coming only every 100,000 years or so, and lasting a few thousand years or tens of thousands of years at most. The Earth is currently in an interglacial period, which has been termed the Holocene or postglacial. The diagram (right) shows only major cycles of cold and warm; there were also many shorter fluctuations. During the coldest episodes, the polar ice sheets expanded to cover areas far to the south of glaciers today, and winter temperatures in England fell to 0°F (–18°C) or lower, similar to those of western Siberia today. The alternation of warm and cold is thought to be due to minor variations in the Earth's orbit.

Temperature scale = Summer sea-surface temperature in the North Atlantic

850,000
700,000
500,000
300,000
100,000
Years ago
Today

°F °C
59 15
50 10
41 5

dated to about 200,000 years ago, and associated with vegetation and fossils of mollusc and insect species typical of a mild climate. Remains (mostly teeth) of a mammoth similar to *M. trogontherii* have also been found in Japan and Taiwan, dating from the interval 1.0–0.7 million years ago, having crossed from the mainland of the Far East. Their teeth developed certain peculiarities, however, and they have been named *Mammuthus protomammonteus*.

## THE WOOLLY MAMMOTH

After the *M. trogontherii* stage, the mammoth's evolution accelerated and, by 150,000 years ago, the true woolly mammoth *Mammuthus primigenius* had appeared in Europe. As in the previous change, the fossil record indicates that the transformation had occurred earlier, and much further east. Remains discovered in the Kolyma region of northeastern Siberia show the transition beginning as much as 750,000 years ago, with fully-evolved woolly mammoths present by

about 400,000 years ago. The continuously severe climate and dry grassland vegetation of this region had promoted the evolution of a species fully tolerant of the cold climate and treeless vegetation of the Ice Age. As before, during a subsequent period of global cooling, the woolly mammoth finally spread south into Europe. By 100,000 years ago, the beginning of the last Ice Age, the mammoths had taken up residence right across the vast expanse of largely treeless vegetation that extended from the British Isles to eastern Siberia (see p. 16).

The environment of the woolly mammoth for much of the last Ice Age was quite rich. Although the climate was generally too cold for trees, clear skies producing longer hours of sunlight and moderate precipitation promoted abundant vegetation, comprising a mixture of the plant species now living in the Arctic tundra with those of a more grassy, steppelike nature. Other inhabitants included now extinct forms, such as the woolly rhinoceros and cave bear, as well as various species still known today, such as reindeer, musk oxen, horses, and bison.

*The world 21,000 years ago, at the peak of the last Ice Age, was very different from that of today. Ice extended in Europe as far south as central England, and in North America to below the Great Lakes. South of the ice there was a vast expanse of meadowlike grassy vegetation, named "mammoth steppe" after its most famous inhabitant, the woolly mammoth. Farther south was a mosaic of parklands and open woods, which in North America was home to the Columbian mammoth* Mammuthus columbi. *Elsewhere, the vegetation was similar to modern steppes, deserts, subtropical savannas, and tropical forests, although their ranges were different from those of today. Mammoths did not live in these areas.*

Continental ice sheets

Mammoth steppe

Parklands and open woodlands

Southern steppe and desert

Subtropical and tropical vegetation

# THE EVOLUTION AND MIGRATION OF MAMMOTHS

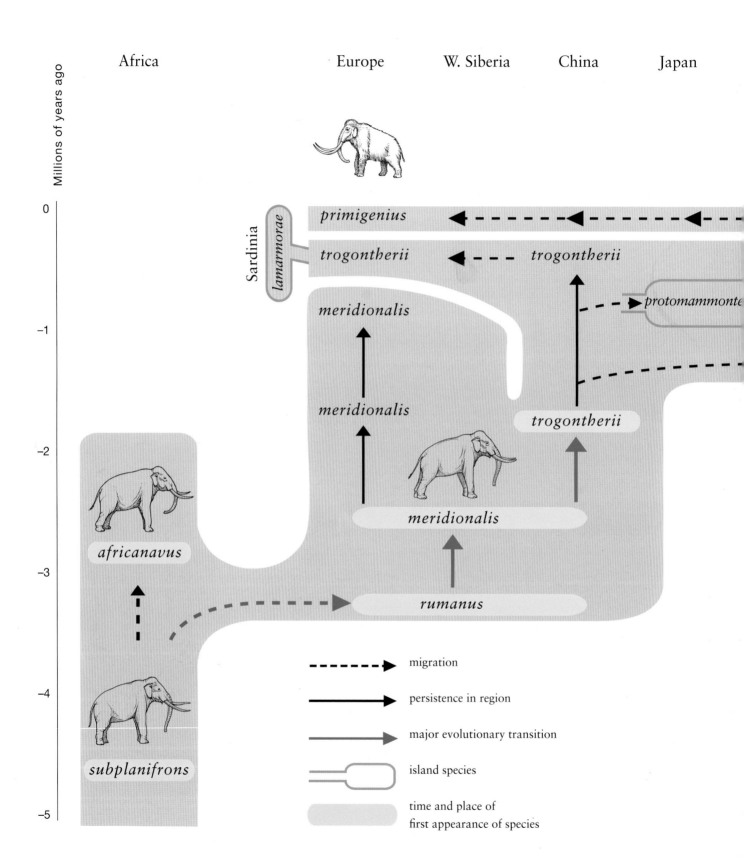

Africa        Europe    W. Siberia    China    Japan

Millions of years ago

0

−1

−2

−3

−4

−5

Sardinia

*lamarmorae*

*primigenius*

*trogontherii*     *trogontherii*

*meridionalis*

*protomammonte*

*meridionalis*

*trogontherii*

*africanavus*

*meridionalis*

*rumanus*

*subplanifrons*

migration

persistence in region

major evolutionary transition

island species

time and place of
first appearance of species

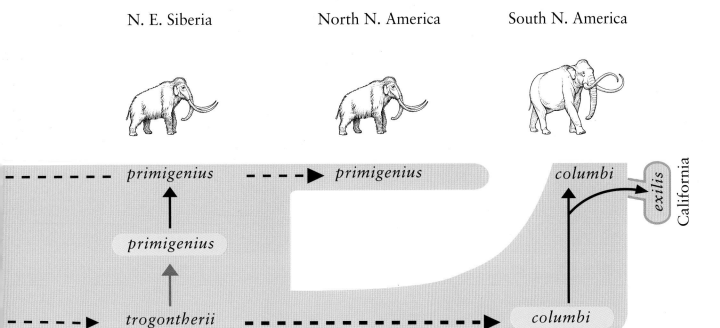

N. E. Siberia          North N. America          South N. America

*primigenius* ---→ *primigenius*          *columbi*          *exilis* California

*primigenius*

*trogontherii* ---→          *columbi*

**Mammoth teeth** *from the Siniaya Balka locality in SW Russia, appear to show the coexistence of M. meridionalis (below, with low tooth crown) and M. trogontherii (above, with high tooth crown) in eastern Europe about 1 million years ago.*

**One of the earliest known fossils** *of wooly mammoth, this molar came to light in deposits around 700,000 years old in the Kolyma basin of northeast Siberia.*

# THE AMERICAN MAMMOTH

Watched by a pair of coyotes, a small family of Columbian mammoths feeds in the rolling landscape of the Black Hills of South Dakota, 25,000 years ago. This species, *Mammuthus columbi*, was exclusive to North America. Its range extended to Mexico—farther south than any other mammoth species since the original migration from Africa millions of years before.

Spread of the American mammoth

*The Columbian mammoth was up to 13 ft (4 m) in height and weighed an estimated 10 tons. The details of its external appearance are in part speculative but, since the animal lived in a more southerly habitat, it is likely that its coat was much less developed than the woolly mammoth's. The tusks, however, were every bit as imposing; those of the females and young shown in the foreground are relatively modest compared to the massive pair carried by full-grown males.*

*In the northern part of its range, the Columbian mammoth may occasionally have encountered its woolly cousin migrating from its more northerly territory. Here, male Columbian and woolly mammoths are confronting each other. The woolly mammoth is distinguished by its smaller size and heavier coat.*

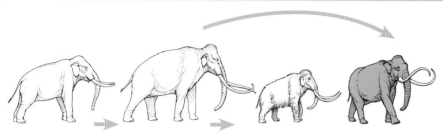

M. meridionalis　　M. trogontherii　　M. primigenius　　M. columbi

*The American branch of the mammoth family,* M. columbi, *may have evolved from early populations of* M. trogontherii *(the steppe mammoth) that entered the New World about 1.5 million years ago. In Eurasia the same ancestor evolved into* M. primigenius, *which much later also migrated into North America.*

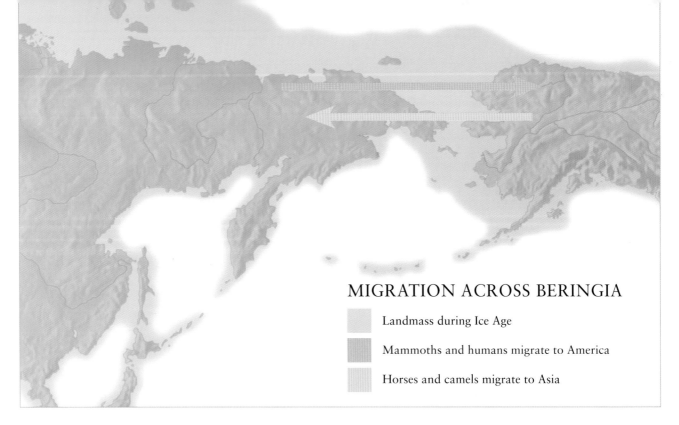

Landmass during Ice Age

Mammoths and humans migrate to America

Horses and camels migrate to Asia

## INTO THE NEW WORLD

Mammoths first spread into North America about 1.5 million years ago. The route to North America was eastward. The northeastern tip of Asia (the Chukotka region of Siberia) and the northwestern tip of America (Alaska) are today separated by 60 miles (100 km) of sea, but during each Ice Age sea levels fell by about 300 ft (100 m) because so much of the world's water was locked up in the ice caps. At such times the Bering Strait between Siberia and Alaska became dry land, forming part of a region known as Beringia, and animals could migrate across this land link between the two continents.

Mammoth remains earlier than about 1.2 million years old are so far unknown from eastern Siberia. However, it is clear that early mammoths must at some time have expanded into Beringia, for this was the only route into North America. The most complete early mammoth skeleton in North America was discovered in 1986 in Anza-Borrego Desert State Park in California. The skeleton, that of an old female, was disarticulated and had clearly been attacked by carnivores and scavengers before burial in a shallow river. The animal had suffered from arthritis of the jaw joint, which probably made chewing difficult and may have contributed to its death. The skeleton has been dated to approximately 1.1 million years ago, and shows that mammoths rapidly dispersed to southern, temperate parts of the continent after their first entry from Siberia.

*At times of low sea level, Asia and America became linked by land. At different times, mammoths, bison, wapiti, and humans spread from the Old World into the New, while camels and horses moved in the opposite direction.*

For a long time it was believed that the earliest mammoths in North America were the ancestral mammoth *M. meridionalis*, since the species was still living in Europe at that time. However, recent research has questioned that assumption. First, it is now known that by 1.5 million years ago, *M. trogontherii* had already evolved in eastern Asia, and this was the region from which the N. American mammoths must have migrated. Second, fossils from Colorado, Florida, and elsewhere, formerly identified as *M. meridionalis*, have been re-studied and found to be of more advanced form. At present, the existence of any mammoth remains as primitive as *M. meridionalis* in North America is uncertain.

Whatever the precise ancestor, the early mammoth immigrants into North America gave rise to a species unique to that continent: the so-called Columbian mammoth, *Mammuthus columbi*. The Columbian mammoth was larger than its woolly cousin and probably did not have such a dense covering of fur. It was adapted to a more open and cooler habitat than *M. meridionalis*, but not to such an extreme degree as the woolly mammoth. In its adaptations to a changing diet it remained similar to its likely direct ancestor *M. trogontherii*, typically with around 20 enamel ridges in its last molars. But whereas in Eurasia *M. trogontherii* evolved farther to produce the woolly mammoth, the Columbian mammoth remained at roughly this level of adaptation until its extinction at the end of the Pleistocene. It occupied much of what is now the United States and Mexico, and even extended as far south as Costa Rica, but did not live in the Arctic regions of Canada, in other words, not as far north as the woolly mammoth.

The Columbian mammoth was part of the rich and varied mammal fauna of the Pleistocene in North America, much of which is now extinct (see pp.146–47). Its remains have been found together with such exotic species as the giant armadillo, sabre-tooth cat, the giant ground sloth, and "yesterday's camel," as well as its distant relative the American mastodon, and large numbers of more familiar animals such as horses and buffalo.

Many American fossils have been named "Imperial Mammoth," *Mammuthus imperator*, and have been thought to represent an earlier stage of evolution, but it now seems likely that these are the same species as the Columbian mammoth. Additionally, some researchers believe that certain populations of the American mammoth lineage, especially in the Midwest, evolved beyond the *M. columbi* stage, producing a more advanced species, *Mammuthus jeffersonii*. However, the existence of *M. jeffersonii* as a species distinct from *M. columbi* is uncertain. The name *jeffersonii* was coined by the American paleontologist Henry Fairfield Osborn, the leading expert on fossil elephants in the early 20th century. He named it after President Thomas Jefferson who, more than a century earlier, had had a particular interest in mammoths and mastodons, which he believed might still be living in the American West.

About 100,000 years ago there was a second wave of mammoth migration, but this time it involved the woolly mammoth

*The lower jaw of the oldest mammoth skeleton in North America, discovered at Anza-Borrego, California, in 1986.*

*M. primigenius*. Like its ancestor *M. trogontherii* more than a million years previously, the woolly mammoth migrated across the dry Bering Strait from the Siberian part of Beringia into North America. Fossils of woolly mammoths dating from this time onward are common in Alaska, Canada, and the northern part of the contiguous United States. The woolly mammoth took up the same range of latitudes and habitats in America as it occupied in the Old World. To the south, the Columbian mammoth was already established, so the two species more or less divided the continent between them.

Although woolly and Columbian mammoths have each been found at hundreds of fossil localities, few sites contain both. However, at Hot Springs in South Dakota (see pp.64–65) some woolly mammoth fossils were found in the same sinkhole as numerous Columbian mammoth skeletons. This site is in the region of overlap between the ranges of the two species. However, it is not clear whether both were there at the same time, or whether woolly mammoths entered the area at a time of cooler climate when the Columbian mammoths had retreated south.

Mammoths never reached South America or southern Asia, nor did they re-enter Africa after the original dispersal. But they came to cover most of Europe, northern Asia, and North America. In the Far East, numerous woolly mammoth localities have been found in north-east China, and during the low sea level the last glaciation, they expanded into the north island (Hokkaido) of Japan. To the south in both China and Japan they were replaced by the woodland elephant *Palaeoloxodon*.

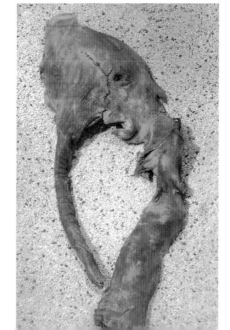

*Although most frozen remains have come from Siberia, the frozen mammoth known as Effie was found in 1948 near Fairbanks, Alaska. This partial carcass of a baby, comprising the head, trunk, and foreleg, graphically illustrates the spread of woolly mammoths from the Old World into the New.*

# MAMMOTHS IN MINIATURE

On several islands around the world, mammoths of greatly reduced body size evolved. Pictured here is *Mammuthus exilis*—a dwarfed version of the Columbian mammoth, *M. columbi*—which lived on the California Channel Islands between 50,000 and 13,000 years ago.

Santa Barbara ● ● Los Angeles

1 3
2

1. San Miguel
2. Santa Rosa
3. Santa Cruz

*During the Ice Age,* sea levels were low enough to connect some of the smaller California Channel Islands, but not to link them to the mainland 6 miles (10 km) or so away, so the original, colonizing mammoths must have arrived by swimming. In an ongoing foot survey of the islands, more than 140 localities with mammoth remains have been found, most of them on Santa Rosa. A complete skeleton was found there in 1994. Some bones of larger mammoths have also been found, but it is uncertain whether these represent stages in a single dwarfing process or successive waves of immigration.

Even on the mainland, mammoths varied in body size between areas and at different times, largely because of the quality of feeding available. None, however, became as small as those stranded on small islands, where the food supply was strictly limited, especially in times of seasonal shortage. Island inhabitants were unable to migrate to richer feeding grounds, so smaller animals that could survive with less food were at an advantage.

A flock of geese, each bird some 2–3 ft (60–90 cm) in height, graphically illustrates the small size of the island mammoths. The California Dwarfs were typically 4–6 ft (1.2–1.8 m) high at the shoulder, less than half the height of their mainland ancestors and contemporaries.

# DWARF MAMMOTHS AND ELEPHANTS

Fossil evidence of dwarf mammoths and elephants has been found in a number of island locations around the world. Caves in Sicily and Malta have yielded many bones of an elephant only 3 ft (1 m) high in the adult, and believed to be at least 500,000 years old. Although their single-domed skull and curved tusks superficially resemble a mammoth, this is a result of dwarfing, and the teeth suggest that the species evolved from the "straight-tusked elephant", *Palaeoloxodon antiquus*, of the European mainland. This extinct species was up to 13 ft (4 m) high and weighed at least 10 tons. The island dwarfs, with an estimated mass of around 100 kg, had therefore been reduced to only one hundredth the body weight of their ancestor. Similar

## "Some adult dwarf elephants were no taller than a goat"

dwarf elephants evolved on Cyprus and at least eight other Mediterranean islands. On one Mediterranean island, Sardinia, the dwarf appears to be a mammoth. A preserved partial skeleton indicates only moderate size reduction from the ancestral mainland mammoth, which may have been *Mammuthus trogontherii*. The Sardinian dwarf has been named *M. lamarmorae*.

On the other side of the world, in the California Channel Islands, a dwarfed species of mammoth appropriately named *Mammuthus exilis* evolved from the American lineage *M. columbi* (see pp.36–37). Until 1994, only fragmentary remains of the dwarf mammoths had been found, but in that year a complete skeleton was excavated from the sediments of an ancient sand dune. The skeleton, 95% complete, represents an aging male, and is radiocarbon dated to around 15,000 years ago. The Californian dwarfs were

typically 4–6 ft (1.2–1.8 m) high at the shoulder, less than half the height of their mainland ancestors and contemporaries. They may have weighed only about 1 ton, compared to the 10 tons of the average Columbian mammoth. In other respects they were probably similar, with short fur and a typical mammoth body form, but a relatively large head. As well as being smaller in size, it has been suggested that the dwarfs had relatively short lower limb bones and a lower centre of gravity, allowing them to climb steeper slopes.

There has not been a land bridge to the Channel Islands in recent geological history, so the founding mammoths must have arrived by swimming from the mainland only 4 miles (6.5 km) away, closer than today because of lowered sea level. The full-sized mammoths probably arrived on the islands some time after 100,000 years ago, perhaps tempted across by the smell of vegetation from the islands. The islands had coalesced into a "super-island" that has been named "Santarosae," four times larger than their combined area today. As sea levels rose again, the islands became once more isolated and smaller, selecting for small size of the animals. Radiocarbon dates on the dwarfs give ages of 15,000–13,000 years ago. Soon after, they were extinct. There is no evidence any of them made the return journey to the mainland.

In 1993 excitement was caused by the announcement of very small woolly mammoths from Wrangel Island in the Arctic Ocean, 125 miles (200 km) north of Siberia. Partially eroded tusks, teeth, and leg bones were found on the surface of the tundra or in shallow stream gravel. Radiocarbon dates indicate that a mammoth population lived there 10,200–3,900 years ago (see p.163). Several very small molar teeth indicate animals just 6 ft (1.8 m) high, although recent discoveries of larger bones suggest that the population may have been unusually variable in size.

*Dwarf mammoths evolved* independently on several islands across the globe. In each case, they were descended from normal-sized animals from the adjacent mainland.

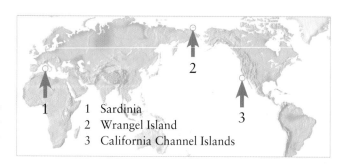

1 Sardinia
2 Wrangel Island
3 California Channel Islands

*Excavation of a mammoth skeleton on Santa Rosa, California Channel Islands, in 1994. The skull and tusk can be seen under the grid on the right; to the left, the pelvic girdle with thigh bone are still articulated.*

Earlier mammoth remains from Wrangel Island, with radiocarbon dates of 13,800 years ago and older, are all of normal, large size. Whether the dwarfs evolved from full-sized mammoths stranded on the island at that time is uncertain. The recent dating of mammoth fossils from the nearby mainland to as late as 10,200–9,600 years ago, raises the possibility that mammoths had died out on Wrangel and re-colonized at that date, just before the final separation of the island from the mainland. These last mainland populations may already have been reduced in size; or else they reduced very rapidly after colonizing the island.

What was the cause of dwarfing in the island mammoths? Most researchers believe that it relates to restricted food resources and the absence of predators. On a limited area, food is at a premium and small-bodied individuals that can survive on less should be favoured by natural selection. This would be particularly felt during winter shortages, when migration to richer feeding grounds is denied to island inhabitants. The absence of wolves and other predators would have added to this process. Small islands cannot hold sufficient numbers of herbivorous mammals to support carnivore populations. In this situation, adaptive reasons for large size—predator avoidance and defence—disappear. Also, with no predators to cull their numbers, herbivore populations expand until they are in direct competition for food—augmenting the pressure for frugality and size reduction.

Recently, some researchers have suggested another mechanism for size reduction. If the pressure was for rapid reproduction, animals would benefit from reaching sexual maturity at a younger age; in other words, they would reach adulthood at a smaller body size, after which they would stop growing.

# DNA FROM MAMMOTHS

The extraction of DNA (genetic material) from fossils represents one of the most exciting developments in the history of paleontology, and the woolly mammoth has received more research attention than any other species. Reports, in the early 1990s, that DNA had been extracted from dinosaur bones proved ill-founded:

*A woolly mammoth jaw* from Allaicha, NE Siberia, shows (right) where bone was cut for DNA analysis.

DNA is a delicate molecule that does not survive for tens of millions of years. Fossils of the mammoth and other Ice-Age animals, however, only tens of thousands of years old, have been found to contain small quantities of DNA, the molecule that codes for the growth and development of all animals. The technology to extract and analyze the tiny, fragmented pieces of DNA has only been available since the late 1980s, but its rapid refinement has produced a burgeoning of research in this exciting area.

The best remains for the preservation of ancient DNA are the frozen carcasses of northern Siberia and Alaska. Here, flesh, skin, and hair may contain DNA as well as many other biological molecules. However, DNA can also be extracted from bones and even dentine from teeth. These body parts, in life, contained living cells whose DNA, after death, has sometimes been preserved in the tiny spaces formerly occupied by the cells. The ingenious processes by which DNA is extracted from the fossils, multiplied up into workable quantities, and then analyzed, are described on p.175. DNA is a chain molecule comprising a sequence of linked smaller molecules called bases. Groups of bases, usually several hundred units long, form the genes. The precise order of the four bases A, C, T, and G in a gene forms the DNA sequence, and determines the animal's structure. The base sequence differs between species and, to a lesser degree, between individuals of a species.

The first mammoth DNA was isolated in 1985, but it was not until 1994 that the base sequence of a short strand of DNA, 375 bases long was determined—about a third of one gene. In this, and many subsequent studies, a particular loop of DNA was targeted—from tiny structures called mitochondria, found within most living cells and containing 37 genes. From the outset, researchers attempted to use the mammoth's DNA structure to determine if it was more similar to that of the Asian or African elephant. The early studies were ambiguous—some genes suggested one answer, some the other. By 2000, lengths of DNA more than 600 bases long had been sequenced, but still the relationship of the mammoth remained elusive. Finally in 2006, using new technology, three research groups independently published the complete mitochondrial DNA sequence of a woolly mammoth—more than 16,000 bases long. This was achieved by sequencing dozens of smaller fragments, and piecing them together based on regions of overlap. The resulting sequence at last appears to answer the long-standing question: the mammoth is more closely related to the Asian elephant, but only just. The likelihood is that the African elephant branched off from the common stock about 6 million years ago, and the Asian elephant and mammoth less than half a million years later. So the genetic differences between each species and the others are all quite similar.

*This small piece* of equipment is at the heart of the process for obtaining DNA sequences from ancient remains. It runs the Polymerase Chain Reaction (PCR), whereby tiny quantities of DNA are multiplied millions of times into quantities sufficient for analysis.

*A DNA **sequencing machine**, where the base sequence is read from PCR products. Each of the four colors represents one of the four "bases" in the DNA chain.*

The information contained within mitochondrial DNA is limited—the vast majority of animal genes reside within the cell's nucleus: over two billion DNA bases. In a paper published in 2005, a revolutionary new technique was used to extract and sequence 13 million bases of mammoth DNA from a Siberian jaw bone. Thousands of small fragments were obtained, and their sequences joined together, using a living elephant's DNA as a template. The potential exists, ultimately, of obtaining the entire genetic sequence (genome) of the mammoth—as has been done for the human and some other animals.

DNA studies promise to yield much information about the mammoth and other Ice-Age animals. First, the study of nuclear genes may allow us to determine aspects of the animal's form and appearance in a way previously considered impossible for extinct species. A first attempt at this approach was announced in 2006: researchers isolated from mammoth DNA a gene known to be involved in the determination of coat color in living mammals. They found different versions of the gene within individuals and between them. These variations corresponded to less and more active pigment production, respectively, and are known in some rodent species to produce lighter and darker hair color. These studies are in their infancy, but they suggest similar variation of hair color among mammoths.

Another potential use of "fossil genetics" is to look for the DNA of disease organisms within the mammoth tissue. In ancient human remains, DNA from tuberculosis and other diseases has been recovered. Studies of mammoth fossils have so far demonstrated the existence of so-called endoviruses: lengths of DNA from viruses that infected mammoths or their ancestors in the distant past, but which had become incorporated in the mammoth's own DNA and were passed harmlessly from generation to generation. Future work may demonstrate which disease organisms actively infected the latest mammoths.

The studies so far described were based on single or a few individuals. Another development in ancient DNA research is to obtain genetic information from hundreds of mammoths across their vast geographical range. In this way we can learn how variable mammoths were, how well they mixed or whether local populations were isolated, and even trace the movements or demise of

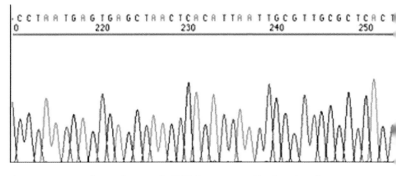

*A **computer readout** of part of a DNA sequence; Each colored "peak" indicates a base (letter of the genetic alphabet); the complete sequence is recorded at the top of the screen.*

individual populations through time. The first study of this kind has shown that Siberian mammoths fell into two rather distinct genetic groups, one of which seems to have died out much earlier (by about 40,000 years ago) than the other (about 12,000 years ago). This in turn may cast light on the process of the final extinction of mammoths (see Ch. 5).

Finally, new techniques are allowing the study of mammoth DNA even in the absence of tangible fossils. Ice and frozen soil in arctic regions have in some cases remained unthawed for thousands of years, and incorporate organic chemicals from the time of their first freezing. Analysis of the thawed "muck" has shown that DNA of numerous species of mammals, insects, plants, fungi, and bacteria can all be preserved in a single sample. In the laboratory, the DNA of the individual species can be isolated and identified. This offers the prospect of detecting the presence of mammoth even in regions where bones are not preserved, and of extending the study of mammoth DNA much further back in time than previously achieved.

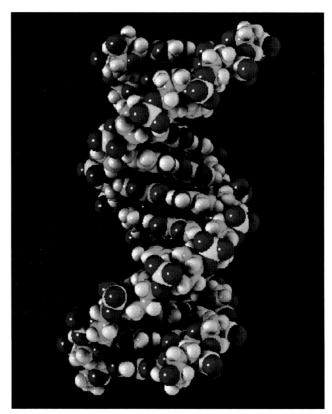

*A molecular model* showing two turns of the double helix of DNA. Only under favorable conditions of temperature and chemical environment can this fragile molecule survive for thousands of years.

*This diagram* illustrates a small part of the DNA sequence of a mammoth, determined for the first time in 1994. Most of the mammoth sequence is identical to that of living elephants, but two differences are highlighted.

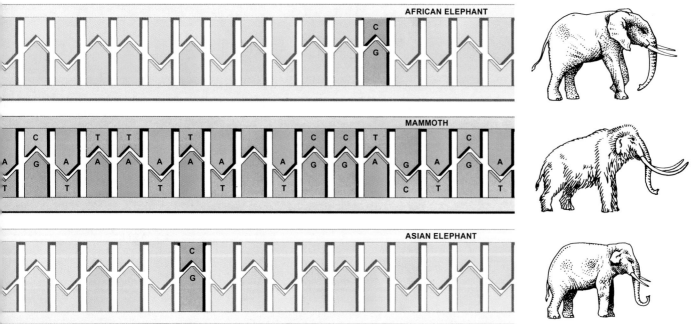

# RESURRECTING THE MAMMOTH?

The finding that DNA is preserved in fossils has inevitably led to speculation that extinct species might be brought back to life. Two distinct methods have been suggested for achieving this feat, though they are often confused.

### Cloning

Cloned laboratory mice were first produced in 1986, and Dolly the sheep was born in 1997. Since then several other domestic animals have been cloned. In order to clone a mammoth, an egg cell (ovum) from a female elephant would be taken, and its nucleus, containing its genetic material in the form of DNA, removed. In its place would be inserted the nucleus of a mammoth cell taken from a frozen carcass. As in cloning of living species, chemical or electrical stimulation might cause the egg to start dividing, at which point it would be placed in the female elephant's uterus, and 22 months later the elephant would give birth to a 100 percent baby mammoth.

### In-vitro fertilisation (IVF)

According to this idea, the elephant's egg cell would not be deprived of its own nucleus, but would be fertilized in the laboratory by a sperm cell taken from the frozen carcass of a male mammoth. Again the egg would be implanted back in the mother elephant, but this time she would give birth to a hybrid baby—half mammoth, half elephant. The process would have to be repeated, crossing the hybrid individual with another mammoth, and so on over several generations until an almost pure mammoth resulted.

There are several serious objections to all such plans:
- Even in the best-preserved mammoth remains, the DNA is fragmented into millions of tiny pieces, and has lost its organization into chromosomes that is essential for the successful development of a baby animal.
- Sperm of living mammals has so far proved to be potent after 15 years of deep-freezing, a thousand times less than the requirement for mammoth IVF.
- In very rare crosses between African and Asian

*The robotic mammoth model at Le Thot, France, moves its ears, eyes, tail, and front legs. It breathes air through its trunk, lifts its head, and trumpets at onlookers.*

elephants, the baby died of defects soon after birth. A mammoth-elephant hybrid would probably suffer the same fate. Even if it survived, the ethics of making such a hybrid "monster" are questionable.
- The natural habitat of the mammoth barely exists anywhere today. Moreover, the mammoth was a highly social animal, so generating one or a few individuals would be less than ideal from a behavioral point of view.
- The scientific time and resources to accomplish such a feat would be enormous, for very uncertain aims or benefits. Such resources would be better deployed in the conservation of the endangered elephants that are still alive.

# MAMMOTHS UNEARTHED

Mammoth remains are by no means rare. Over the centuries they have been found on four continents and in a wide variety of locations: in the frozen wastes of Siberia and Alaska, in the tarpits of Los Angeles, on the bed of the North Sea, in gravel pits, caves, and coal mines. Their bodies have been revealed in every possible condition, and range from specimens with hair, hide, flesh, blood, and internal organs to skeletons and mere fragments of teeth and bones. They have been found by all kinds of people—ivory collectors, hunters, miners, bulldozer drivers, boat crews, and people out walking the dog.

Such discoveries were once a source of dread and superstition, but gradually scientists have been able to use these finds to build up a thorough picture of this great creature of the past. By carefully piecing together the evidence presented by mammoth finds it has been possible to learn about the animal's life, its numerous causes of death, and the conditions leading to its varied states of preservation.

*A mammoth skeleton is carefully excavated* at the Hot Springs site in South Dakota, *where up to 100 mammoths met their deaths in a sinkhole some 30,000 years ago. Excavations proceed every July and can be observed by visitors to the site.*

# DISCOVERY OF THE CENTURY

On 15 May 2007, a nomadic reindeer tribesman by the name of Yuri Khudi spotted parts of an animal's body sticking out of damp snow. Thinking at first that it was a dead reindeer, he went over to investigate. Khudi, a Nenets, had stumbled across arguably the most complete carcass of a woolly mammoth, or indeed of any prehistoric animal, ever discovered.

Recognizing what he had uncovered, he traveled on foot to the nearest village, and then by helicopter to alert the regional museum to his find. The mammoth, 51 in (130 cm) long and 35 in (90 cm) tall was a calf, probably less than a year old. The preserved genitalia indicate that she was a female, and she has been named Lyuba in honor of the finder's wife. At around 110 lb (50 kg), her weight is probably about half that in life; like other frozen mammoth remains, she was preserved by a process akin to mummification, by loss of water to the surrounding icy sediment (see p.50).

Lyuba's preservation is exceptional, retaining the body's full three-dimensional shape—unlike the flattened carcass of Dima, the baby mammoth discovered in 1977 (see p. 60–61). The entire skin covering is intact and in excellent condition, promising good preservation of the internal organs. Much of the fur has been lost, but tufts of brown hair are preserved here and there on the body.

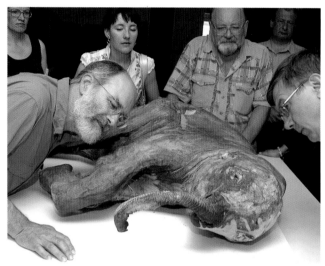

*Scientists gather to examine the carcass, especially the double tip of the trunk. The flanges on the lower part of the trunk, and the form of its double tip, are unique to mammoths.*

## "Lyuba's preservation is exceptional, retaining the body's full three-dimensional shape"

The carcass was discovered on the bank of the Yuribei river in the Yamal peninsula of northern Siberia —at the westernmost edge of the range across which frozen mammoth carcasses have been discovered. The lack of damage or decay suggests that the animal either died by sinking into soft ground, or else was buried almost immediately after death. Freezing must have been rapid and the area subsequently covered with sediment, ensuring that the carcass remained in the permafrost layer—protected from seasonal thawing that would have rapidly ensured decay. After millennia of burial, soil erosion finally exposed the carcass. The moments leading to the animal's death can only be surmised. Baby mammoths were undoubtedly like living elephant calves in sticking close to their mother at all times. Occasionally they might venture a few

yards away, before scurrying back to the safety of the group. Even if the mother had died, other females in the herd would have taken over the maternal role. A likely scenario, therefore, is that the Yamal baby fell suddenly into a crevasse or mud pool. In that situation, the females of the herd would have done everything in their power to haul her out. In this case they failed.

Only one part is missing from the otherwise complete carcass: the tail, apparently removed—perhaps by a fox —after exposure. Yuri Khudi's discovery of the young mammoth graphically illustrates the chance nature of many of the most important fossil finds; weeks or even days later, the carcass would have been badly damaged or gone completely.

*The Yamal baby mammoth carcass, discovered in 2007. Exceptional in its preservation, the young animal must have been frozen very soon after death and has remained in a natural "deep-freeze" ever since. Discovered accidentally by a reindeer-herder in northern Siberia, she is the subject of intensive scientific research and is likely to yield new information about the life and death of the woolly mammoth.*

Until only a few centuries ago, any large bones discovered in the fields or caves of Europe were usually assumed to be the remains of giants, and were often displayed as curiosities in castles, palaces, town halls, churches, and monasteries. Among the best known is a mammoth thigh bone—engraved with its date of discovery (1443) and the motto of Emperor Frederick III—which for about three centuries hung in the "Giant's Portal" of Vienna's Cathedral of St. Stephen. In a church in Valencia, a mammoth tooth was venerated as a relic of St. Christopher, and a mammoth thigh bone was carried in triumph through the streets as a saint's arm when special prayers were needed to bring rain.

In 1577 a collection of mammoth bones sparked a heated theological debate in Switzerland. When it was proposed that they should be given a Christian burial in a cemetery, a doctor studied them closely and declared them to be the bones of a giant, 18 ft (5.5 m) tall, who deserved no such courtesy, being a heathen. It is even possible that fossil elephant skulls lie behind the legend of the Cyclops of Greek mythology, since their large nasal opening might have been seen as the giant's single eye-socket.

## "In the distant past mammoth bones were assumed to be the remains of giants"

Eventually, the true nature of such remains dawned on a few scholars. For example, in 1613, some bones found near Lyon, France, were attributed to the giant Teutobocus, king of the Teutons in 105 BC, but a medical student called Riolan correctly attributed them to an elephant. In 1696 a skeleton found at Touna, in the Grand Duchy of Gotha, Germany, was declared by the Gotha medical college to be an elephant-shaped freak of nature, the result of a mysterious playful force that produced all kinds of animal shapes in stone—but a historian to the dukes of Saxony examined it and correctly identified it as that of an extinct elephant.

This interpretation, however, presented a difficulty: if such remains belonged to elephants, which today live in warm southern countries, how could they possibly be found in northern Europe? Until the mid-18th century any finds in Italy were attributed to animals brought north in Roman times: for example, it was known that

Hannibal, as well as Pyrrhus, king of Epirus (in modern Greece), had used elephants against the Romans. But with more northerly finds scholars could only assume that the climate in these countries used to be warmer than it is now, or that drowned elephants had been carried north by the biblical Flood.

In 1796 the great French anatomist Georges Cuvier put forward the proposition, on the basis of comparative anatomy, that the remains were those of fossil elephants similar to, but quite distinct from, those existing today. The mammoth was in fact among the first extinct animals to be discovered and studied. In 1799 it was given a scientific name by J. F. Blumenbach: *Elephas primigenius*.

Meanwhile, in North America, numerous bones believed to belong to elephants (actually those of mastodons) had been found in the 18th century in the salty bog soil of Big Bone Lick in Kentucky. A Frenchman, Baron de Longueuil, acquired a tusk, several molars, and a thigh bone, and shipped them back to France. At this time American scholars followed Thomas Jefferson in believing that there was no such thing as an extinct animal: if the bones existed, then so must the living animal. This is reflected in the first name given to the species: *incognitum* (unknown). It was

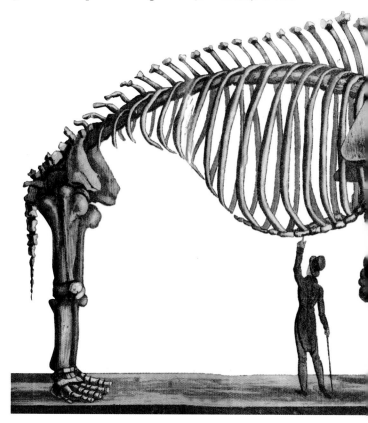

Cuvier, once again, who demonstrated that although the mastodon was related to the elephant, it was quite distinct from both the African and Asian forms and from the Eurasian mammoth.

Mammoth ivory had been excavated and traded by local people right across northern Siberia long before the Russians explored these lands. Markets of the Orient and Middle East (China, Mongolia, and Iran) were acquainted with fossil mammoth bones long before Europeans.

The first accounts to reach Europe from Siberia in the 17th century were very vague, concerning an enormous mammal that still lived on the frozen coasts of Tartaria, and that was identified with the behemoth of the Bible (see box below): it is likely that such reports referred not only to frozen mammoths but also to living walruses and sea cows.

## WHAT'S IN A NAME?

The origins and meaning of the word mammoth remain obscure. Some claim that it is taken from the Arabic *mehemot*, derived from the Hebrew *behemoth*, the name for a primeval creature mentioned in the Old Testament of the Bible (Job 40: 15–24). An animal of enormous size and strength, with curved horns or tusks, the behemoth was believed to consume huge amounts of grass and water, and had a mild and peaceful nature. An alternative view traces the name to various northern languages.

It is often thought to be of Tartar origin, but no such word exists in any Siberian tongue: the Tungus call the animal *cheli*, and the Yakuts call it *uukyla*. One possible source is the western Tungus term *namendi* (bear), but the most plausible candidate is Estonian, in which *maa* means earth and *mutt* means mole, linking the name to the widespread belief that the animal burrowed beneath the ground (see p.55).

Whatever its origin, the word appears to have been introduced into Europe in 1618–20 by Richard Johnson, who reported on the maimanto tusks found by the Samoyeds of Siberia. By the 18th century the term "mammoth," in its various forms, had spread into virtually every European language.

*The first attempts to draw a mammoth on the basis of its bones had led to strange unicorn-like beasts (as in Leibnitz's Protogea of the 17th century) or, in the 18th century, to an anomalous ox with hoofs and a single great horn on its forehead. Tusks were also sometimes seen as the talons of a vast bird, the antlers of a giant deer, or the horns of a huge male goat. Even after the existence of extinct fossil elephants was established, it was by no means always straightforward to reconstruct their appearance, and tusks were sometimes assumed to point downwards rather than up.*

## SIBERIAN CARCASSES: MAMMOTHS IN THE FREEZER

The bodies of animals—primarily mammoths, but also other species such as woolly rhinoceroses—have been emerging from the Siberian permafrost since the end of the Ice Age. Siberians and Inuit believed the mammoth to be a living animal because, on seeing its remains—including flesh and blood—exposed by rivers and thawing, they assumed it was some kind of gigantic mole that occasionally came to the surface like a whale, but which died immediately on exposure to sun- or moonlight. This explained neatly why nobody ever saw one alive.

The distribution of these frozen remains, unsurprisingly, corresponds to that of the permafrost, nature's deep freeze; hence they are found only north of 60° latitude, and mostly above the Arctic Circle. Ever since the late Pleistocene (and in some places for at least a million years), the ground here has been frozen to a depth of up to 1,500 ft (500 m), and in the brief summer only the top 5 ft (1.5 m) thaw out.

No single catastrophic event can account for all these remains and there is no real evidence that any of the animals slowly froze to death: many specimens appear healthy, with full stomachs. Some clearly died of asphyxia, either by drowning or by being buried

*Many mammoth remains are found in yedoma, rounded hills of silt composed of up to 80 percent ice. Warm, rainy weather causes them to erode rapidly, making it difficult to stop the frozen site thawing before the arrival of mammoth specialists.*

alive in a mudflow or when the ground above them caved in. Some probably got bogged down in marshy places, while others may have crashed through thin river ice or into concealed, snow-filled gullies.

Such accidental deaths and burials may explain why the Siberian carcasses are predominantly mammoths and rhinos: these were heavy-footed giants, whose sheer size would make it especially dangerous for them to graze at the soft edges of a gully or river bluff, and hard for them to extricate themselves when trapped. Often the animals, alive or dead, were then enveloped by solifluction—water-saturated sediments that slid downhill, then froze around them.

It is the movement of ice into the sediments around the bodies that accounts for the preservation of these "mummies:" over time the carcasses dehydrated as the moisture was drawn into the surrounding layers of ice and crystallized. However, by no means all bodies are intact in the permafrost, since many animals seem to have been exposed to predators and the elements for some time before being buried and, therefore, had largely decomposed prior to preservation.

MAMMOTHS UNEARTHED

Most of the preserved carcasses in Siberia are dated to two periods: before 30,000 years ago, and between 15,500 and 11,500 years ago. The intervening millennia have yielded mostly skeletal material. One possible explanation for this is that these periods had a slightly milder climate (the coldest period was 30,000 to 18,000 years ago). With more water available to create mudflows, carcasses would be more likely to become covered and hence have been more effectively preserved.

The earliest recorded mention of a frozen mammoth is that of Nikolaus Witsen, a Dutchman, who in 1692 wrote of his travels through Siberia and mentioned that mammoths were dark-brown and stank. In 1707, another Dutchman, Evert Ysbrand Ides, crossed Siberia with an Imperial Embassy to the Emperor of China, and was told by an ivory collector that he had once found a huge elephant's head in a fallen chunk of frozen earth, and a frozen foot nearby, probably in the Yenisei region; and, on another occasion, a pair of tusks weighing a total of 432 lb (196 kg).

By the early 18th century Tsar Peter the Great was taking an interest in mammoth bones: he personally examined some of the specimens from Kostenki, and thought them the remains of war elephants from a wandering army of ancient Greeks. In 1722 he put out a decree for the governor of Siberia to find a full carcass and send it to St. Petersburg. An expedition headed by D. Messerschmidt returned to the capital with the skull and bones of a mammoth from the Indigirka river in northeast Siberia. However, the desire for a complete carcass was not to be satisfied for almost a century.

## THE ADAMS MAMMOTH

In 1799 Ossip Shumakhov, a Tungus ivory collector and hunter, noticed a shapeless mass among the blocks of ice at the edge of a swamp by the mouth of the Lena River. The following year it had become further disengaged from the earth, and by the end of 1801 one entire side and a tusk of a mammoth were visible. His wife and friends believed it to be an omen of calamity, since the old men had heard that a similar find had once led to the death of the discoverer's whole family. Shumakhov nonetheless determined to profit from his find by selling the enormous tusks. By the end of 1803, the huge mass of this intact mammoth had fallen onto a bank of sand, and the following year Shumakhov cut off the tusks and sold them to a merchant for 50 roubles.

In 1806 the site was visited by Mikhail Adams, a botanist attached to the Russian Academy of Sciences, who found the mammoth still in situ, although decomposed and mutilated. Shumakhov claimed that it had once been so fat that its belly hung down below its knees. The local Yakuts had fed some of its flesh to their dogs, and the carcass was surrounded by tracks of bears, wolves, wolverines, and foxes. As a result, the skeleton was almost entirely defleshed, but was still complete except for one foreleg. Adams carefully gathered up the skeleton, as well as large quantities of hair and skin—in particular, the dry skin of the side of the head, and the wrinkled right ear with a tuft of hair (see p.87). The brain had dried up, and the trunk was missing. The skin was dark gray, and covered with reddish wool and coarse long black hair. The skin from the side on which the animal had been lying was well preserved, and ten people had great difficulty in carrying it to shore. More than 36 lb (16.5 kg) of hair were recovered from the ground. Everything was sent to St. Petersburg, including the curving 10-ft (3-m) tusks, which Adams had managed to repurchase.

In 1808 the Adams mammoth skeleton was reassembled, with the few missing bones replaced by plaster and wood copies. It was the first ever mammoth skeleton to be mounted. Over 16 ft 6 in (5 m) long and 10 ft (3 m) tall, it was a male which had died at the age of about 45 years.

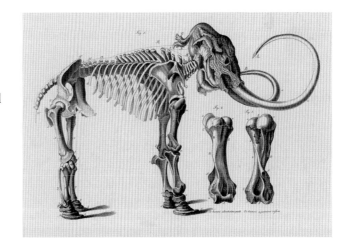

*The skeleton of the Adams mammoth* was reassembled in St. Petersburg in 1808, and samples of its skin and hair were sent to many European and American museums. The location of the find and the animal's tooth structure led Cuvier to his pioneering conclusion that it was a cold-adapted, extinct, local species rather than a victim of the biblical Flood.

# THE BERESOVKA MAMMOTH

An expedition which set out from St. Petersburg in 1901, led by Dr. Otto Herz and Eugen Pfizenmayer, zoologists from the Imperial Academy of Sciences, was responsible for the most thoroughly documented recovery of a frozen mammoth. The previous year the governor of Yakutsk had reported the discovery of a mammoth in an almost perfect state of preservation, frozen in a cliff along the Beresovka River, a tributary of the Kolyma, 745 miles (1,200 km) west of the Bering Strait and 62 miles (100 km) inside the Arctic Circle. A Lamut deer hunter had first spotted a huge mammoth tusk, weighing about 175 lb (80 kg), and then, below it, the head of a second mammoth protruding from the ground, with one smaller tusk of about 65 lb (30 kg).

The hunter sold the two tusks in Kolyma, but did nothing else to the carcass on account of Lamut superstition about such mammoth finds. The buyer reported the find and the Russian finance minister assigned 16,300 roubles for an expedition to examine and secure the specimen. It reached the mammoth after a four-month journey (see box on opposite page). Pfizenmayer gave a graphic description of their first impressions: "Some time before the mammoth body came in view I smelt its anything but pleasant odor—like the smell of a badly kept stable heavily blended with that of offal. Then, around a bend in the path, a towering skull appeared, and we stood at the grave of the diluvial monster...We stood speechless in front of this evidence of the prehistoric world, which had been preserved almost intact in its grave of ice throughout the ages. For long we could not tear ourselves away from this primeval creature, so hung about with legend, the mere sight of which fills the simple children of the woods and tundras with superstitious dread".

The mammoth was later identified as a male, aged about 35 to 40, and about 35,000 years old. Herz and Pfizenmayer believed it had died in this position, on the spot, having probably broken through a thin layer of earth into an ice fissure. Falls of earth had buried and suffocated it. Death by asphyxiation was indicated by the erect penis, which was 34 in (86 cm) long and 7 in (18 cm) in diameter. The animal still had food in its mouth, confirming its rapid demise (see Ch.3). The pelvis, right shoulder blade, and several ribs were broken, suggesting to Herz that it had fallen with great violence into a crevasse. Later investigators came to the conclusion that the mammoth had not died in situ, however, but had moved within a landslide that had resulted from the thawing of ice under the tundra. This might account for the broken bones. The death of the Beresovka mammoth probably occurred in the fall, when the surface soil was still mobile after the summer thaw, but temperatures were dropping, allowing rapid freezing of the buried mammoth.

The carcass had to be cut into pieces because it could not be thawed intact from the frozen ground, nor could its great bulk have been transported whole to St. Petersburg. The job of recovering the carcass was made particularly unpleasant by the stench. However, it appealed greatly to animals—in fact the mammoth had been discovered because the Lamut hunter's dog had been enticed by the smell.

*Most of the head skin and the trunk had been eaten by bears and wolves. The left foreleg was still bent, as if the animal had attempted to lever itself out of the crevasse.*

*About a third of a stuffed replica, displayed in St. Petersburg's zoological museum, is covered with the original skin and hair of the Beresovka mammoth.*

# THE JOURNEY TO A MAMMOTH

On May 3 the Herz–Pfizenmayer expedition left St. Petersburg. It was to cross more than one-third of the Earth's circumference, without leaving the Tsar's empire. The first stage was the easiest, traveling by train to Irkutsk, which was reached on May 14. From Irkutsk, the party proceeded by carts and boats to Yakutsk, which was reached on June 14. The next 2,000 miles (3,200 km) was accomplished on foot and on horseback. At Srednaia Kolymsk the expedition had to stock up with provisions: not only food but also tools to break frozen ground, mosquito nets, gloves, and a collapsible boat which could also serve as a tent. A guide and interpreter were also hired. The party finally reached the Beresovka mammoth on September 9, after a journey of four months.

Dr. Otto Herz and Eugen Pfizenmayer removed about 287 lb (130 kg) of flesh from around the mammoth's hindquarters; this was wrapped in cow and horse hides and allowed to freeze again in the open to preserve it. Skin was cut from the abdomen—507 lb (230 kg) of it—as well as from the head, which included the cheeks, right eyelid, and lips.

Most of the internal organs had rotted away before freezing could preserve them. The stomach was badly decayed and torn, but 26 lb (12 kg) of food fragments were recovered, confirming that the animal did not die of hunger. The dissection of the Beresovka mammoth was completed by October 11, and the remains were packed into 27 cases and placed on 10 sleighs, drawn by reindeer. During the long, arduous journey back to Yakutsk the temperature sometimes descended to −54°F (−48°C); sleighs disintegrated on the unsurfaced roads, and reindeer broke legs on hidden tree stumps. For the 1,865-mile (3,000-km) journey from Yakutsk to Irkutsk horses were used. The party finally reached St. Petersburg on February 18—after just under 10 months of travel by rail and boat, 4,000 miles (6,400 km) by sleigh, 2,000 miles (3,200 km) on horseback, and six weeks of excavation in the harshest conditions.

*A timber structure was built around the carcass, which was still frozen since there was a layer of ice underneath it. Two stoves were used to accelerate the thawing. Herz stands on the left, with Pfizenmayer next to him. A suitable emblem was designed for the flag that flies overhead.*

The mammoth's body was reassembled in St. Petersburg. Tsar Nicholas II and his wife Alexandra visited the still-reeking specimen: the Tsar listened to the explanations with interest, while his wife stood with her handkerchief pressed to her nose.

# LEGENDS ABOUT EATING MAMMOTH FLESH

Many legends surround the excavation of frozen carcasses, and of these some of the more lurid relate to claims of feasting on mammoth steak. The flesh of the Beresovka mammoth (see pp.52–53), streaked and marbled with thick layers of fat, looked quite fresh and healthy as long as it was frozen, and was dark red like frozen beef or horse meat. Herz, the expedition leader, remarked, "It looked so appetizing that we wondered for some time whether we should not taste it, but no one would venture to take it into his mouth...the dogs cleaned up whatever mammoth meat was thrown to them." They were probably wise to abstain since, on thawing, the flesh turned gray. One tale recounted that a scientist did in fact take a bite of the Beresovka mammoth, but could not keep the meat down. There are even accounts of mammoth banquets held in St. Petersburg. Another episode concerns an English lord, Talbot Clifton, who undertook a hunting expedition to Siberia in 1901. On the train to Yakutsk, he chanced on Herz and Pfizenmayer. As his wife, Violet Clifton, later recorded, "Herz counseled Talbot to cast his lot with him and join him in search of a mammoth. Talbot was a little tempted to join the man of science, but he did not assent, for he had the instinct to be alone on his travels, not answerable to anyone." Pfizenmayer, however, had a different recollection: "We had got to know an Englishman, who told us he was traveling... to hunt the local wild sheep. When he heard that we were going there too, only much farther east, he was very anxious to join us. His lordship did not seem to see that we were not at all keen on this."

So Talbot and the expedition went their separate ways, but on Christmas Day 1901, Lord Clifton was back in Yakutsk. "As a gift Professor Herz had sent some of the flesh of the mammoth that he had found. They ate it thoughtfully, for was it not about eight thousand years old?" However, the reliability of Clifton's accounts is perhaps open to question. In China traders recommended mammoth meat as cooling and wholesome to eat, and a remedy for fever. One Chinese text of 1712 states, "Its meat is chilly and cold....Taking it as food, uneasiness and fever can be ridded of and its Russian name is *momentuowa*."

In the 1920s there were several reported cases of travelers being offered, and eating, mammoth flesh. The Chukchi of northeastern Siberia apparently considered mammoths to be evil spirits and had to utter incantations and beat drums whenever tusks were noticed protruding from the earth; but there is a legend that once, when a whole carcass appeared, two Chukchi ate the meat, found it very nutritious and lived on it all winter. Some modern Siberian hunters claim to have eaten cooked mammoth meat and describe it as very fibrous and tough.

*The lower jaw* of the Berovska mammoth, with protruding tongue. The finding of the Beresovka carcass in Siberia in 1900 led to several stories *about the consumption of mammoth meat.*

# LEGENDS ABOUT LIVING MAMMOTHS

Various legends told by peoples of northern latitudes preserve the idea of a living mammoth. For example, the Inuit of the Bering Strait believed in a huge animal that burrowed underground. Some Alaskans said that "Kilukpuk" had lived in the sea with the other big cetaceans until it quarreled with Aglu, another sea monster, and was kicked out and forced to live on land. Kilukpuk therefore "swam" under the ground, and had teeth like a walrus, but straight. Similar stories could be found all over northern Asia: the Ostiaks, Tungus, and Yakuts imagined a huge lemming-like rodent that lived underground and whose movements were responsible for earthquakes. Such tales have even been found in South America: in the 19th century Charles Darwin's guides showed him a mastodon skeleton on the Paran River in Argentina, and claimed it to be a burrower of enormous size. Some old Yakuts still believe that the mammoth is a living animal that burrows beneath the ground. They will cover up any exposed remains to avoid illness and bad fortune.

***Extravagant tales*** *surrounding the discovery of frozen Siberian mammoths were the source of legends about living mammoths, and likewise fueled the imagination of 19th-century illustrators.*

Many North American Indian tribes also had tales that seem to concern the animal: the Northeast Algonkians told of a "great moose" with a kind of limb growing between its shoulders, a fifth leg used to prepare its bed; while the Naskapi of northeastern Labrador knew of a monster with a long nose which it used to hit people.

The Chinese never associated the quantities of fossil ivory they imported from Siberia with living elephants, but believed that the tusks were dragon bones, or the teeth of a giant rodent "Tien-shu," which was the size of an ox or buffalo. The Chinese text of 1712 states that: "The northern plain near the sea in Russia is the coldest place. There is a kind of beast, which like a mouse as big as an elephant, crawls in tunnels, and dies as it meets the sun or moon light. The native people often find it near the river bank." Apparent sightings of mammoths in Siberia have continued into more recent times. At the end of the 16th century, beyond the Ural Mountains, a traveler reported meeting a "large hairy elephant," which the natives described as a valued source of food. In 1920 a hunter, after four years alone in the taiga, stated that he had come upon huge, egg-shaped tracks about 2 ft (60 cm) long. He also claimed to have encountered a large heap of dung containing vegetation, and tree branches broken at a height of 10 ft (3 m). He followed the tracks for days; eventually he saw an enormous elephant, with white, very curved tusks. The animal was dark brown and had long hair.

Even in these instances, other explanations seem more likely. Siberian and Alaskan natives have been digging up mammoth carcasses for centuries, and this could be the source of their knowledge of the animal's appearance and legends about seeing the animal alive. While one cannot totally rule out the possibility of more recent survival (Siberia covers millions of square miles and is only sparsely populated), present evidence indicates that mammoths died out thousands of years ago (see Ch.5).

## LATER SIBERIAN FINDS

Between 1901 and 1903 a Russian expedition to the Liakhov Islands recovered a small adult male mammoth, 8 ft (2.4 m) tall and 14 ft (4.25 m) long. It was presented to the Jardin des Plantes in Paris in 1914, shortly before an edict forbade any piece of mammoth to leave Russia permanently; it is thus the only full Siberian skeleton outside its native land.

In 1908 Eugen Pfizenmayer set out to recover a mammoth on the Sanga-Yurakh River, in Yakutia. He reported that the locals were astonished to find him in good health, in view of the legends that anyone who disturbed a mammoth could not escape misfortune. The earlier death of Herz had confirmed their suspicions. The Sanga-Yurakh specimen proved to be a disappointment since most of it had been eaten by foxes and both tusks were missing. However, its trunk was almost complete: no trunk had ever been recovered from earlier finds. The animal was a female about 60 years old and is thought to have died in winter at least 30,000 years ago when it became stuck in silt on the river bank.

The next major find was that of the Taimyr mammoth, discovered in 1948 by geologists in permafrosted peaty soil in the valley of the Mammoth River, in the northwest Taimyr Peninsula. A small adult male, aged about 50–55 and dated to about 13,500 years ago, it was found with some soft tissue, skin, and hair. The skeleton is remarkably complete—almost all of the bones were recovered except for a few from the

*The Shandrin mammoth, found embedded in loam and gravel, is at least 45,000 years old. Unfortunately, it was washed out with a motorized pump which pulverized the chest organs and destroyed any intact pieces of skin on the ribs. The intestines, however, formed a well-preserved mass in the permafrost.*

toes and tail. The Taimyr mammoth was, therefore, defined in 1990 as a standard of comparison for the species *Mammuthus primigenius*, the woolly mammoth.

In the summer of 1972 a mammoth was excavated on the bank of the Shandrin River, a tributary of the Indigirka. It turned out to be the almost complete skeleton of an old but not particularly large male dating to at least 45,000 years ago (see above). The animal's position, lying on its stomach with its legs pointing forward, is characteristic of dying elephants, so this specimen may simply have had a natural death, probably in the late summer. Its great importance lies in its viscera: the gastrointestinal tract was cut into segments and enabled scientists to reconstruct for the very first time the abdominal organs of a mammoth—the spleen, pancreas, kidneys, intestines, and so on. The stomach contained 641 lb (291 kg) of plant material (see pp.90–91).

In 1977, the year of Dima's discovery (see pp.60–61), a small female mammoth 10–14 years old was found in a hill on the right bank of the Yuribei River, on the Gydanskij Peninsula, while a third specimen, the Khatanga mammoth,

*The Yukagir mammoth, discovered in 2002, was the first to be studied by computer tomography.*

was discovered by a reindeer herder in alluvial sands on the left bank of the Bolshaya Rassokha River. It was an adult male, about 40 years old, and became one of the first mammoths from which DNA (genetic material) was successfully extracted (see pp.40–43). The partly decayed trunk was reported to have had pink-coloured skin.

A mammoth calf, since named Mascha, found in 1988 on the Yamal Peninsula, is the westernmost frozen specimen yet discovered. The sighting, by the commander of a ship, was a complete fluke, as damage to the vessel caused it to stop precisely at the spot where the carcass lay. This baby female, aged only 3–4 months, was probably washed out of a frozen deposit during the last Ice Age, then carried away by a flood before being refrozen.

A further mammoth calf, apparently of even younger age, was discovered in 1990, and excavated in 1991 at Mylykhchan on the Indigirka river. The hair on its head, at least as preserved, was bright yellow.

*In the spring of 1998, French explorer Bernard Buigues, traveling in the Taimyr peninsula of northern Siberia, was led by native hunters to a remote spot on the tundra where two huge, perfectly preserved mammoth tusks had been found.*

Recent years have seen new techniques applied to the discovery and excavation of mammoth carcasses. In 1997, the discovery of the Jarkov mammoth, on the Taimyr peninsula, attracted widespread media coverage. Two huge, perfectly preserved mammoth tusks (see below) had been found by local people. Further excavation revealed the skull, from which all flesh had disappeared, but it was realized that this had resulted from its proximity to the surface, where the soil thaws for a few months each year. The rest of the carcass might be perfectly preserved deeper in the permafrost. Instead of attempting excavation under the harsh conditions of the site, heavy machinery was used to cut out a huge, eight-cubic-meter block of frozen sediment corresponding to the expected position of the carcass. The cube, weighing 23 tons, was airlifted by helicopter to an underground laboratory in Khatanga, where it is held in naturally sub-zero temperatures for gradual preparation and scientific study. The male carcass, approximately 24,000 years old, is not complete, but includes skeletal remains, some flesh, and quantities of hair.

A further development has been the use of ground-penetrating radar (GPR) for the location of fossil remains. The technique was used to good effect in

the recovery of another carcass from the Taimyr, the Fishhook Mammoth, in 2001. The portable equipment accurately located bones and tissue through the upper few meters of permafrost. The Fishhook mammoth, an aged male of similar geological age to the Jarkov animal, included some intestinal remains.

The most important frozen finds of recent years have been made on the banks of the Maxunuokha River in northern Yakutia, Arctic Siberia. In 2002–4 an exceptionally well-preserved head and foreleg were excavated, as well as the vertebral column, rib cage, and intestinal contents. These remains, known as the Yukagir mammoth, have added significantly to our understanding of mammoth anatomy and diet (see p.56 and Ch.3). The specimen, dated to about 22,500 years ago, was seen by an estimated 6.6 million people when it was exhibited at EXPO 2005 in Japan.

New finds continue to be made. The skin of a mammoth's hindquarters, with perfectly-preserved tail, was found in 2003 on Bolshoi Lyakhowski Island (see Ch.3). In 2004, the head and front part of a mammoth calf were found in a gold mine in the Oimyakon district of Yakutia. In the same year, a nearly complete skeleton, plus a forefoot with skin and

*Only a handful of the mammoths found in Siberia are as complete as Mascha, seen here lying where she was found. The carcass was spotted by a passing boat on the bank of a tributary of the River Ob. A laceration to the back of her right hind foot—with shredded, blackened fibers of sinew— may have been the cause of death.*

hair, and the contents of the animal's stomach, were discovered in the basin of the Mongocheyaha River on the Gydansk peninsula of north central Siberia.

The total of frozen mammoth finds known from the whole of Siberia was reckoned in 1933 to be only 34 for the previous 225 years, and a mere handful of those were complete, the rest being fragments or skeletons. There are innumerable creeks and lakes in the vast area of northern Siberia, so it is almost certain that hundreds, if not thousands, of frozen mammoths have emerged from their shores over time and continue to do so; but the region has always been very sparsely populated, and before gold-mining operations began the only discoveries were chance finds made by hunters, fishermen, herders, or ivory collectors. Countless specimens must have been destroyed by animals, putrefaction or floodwaters without ever being seen by human eye. Indeed, even hunters can only have

# WANTED MAMMOTHS

Mammoth remains were unearthed sporadically throughout the 19th century, and in 1860 the Academy of Sciences produced a leaflet that offered a reward of 100 roubles to anyone who discovered and reported a complete skeleton, and an additional 50 roubles if the Academy was satisfied with the find. Discoverers could sell the tusks to anyone they wished. If the meat and hide were present, the reward was to be 300 roubles (later increased to 1,000).

However, many local Siberians remained unenthusiastic, feeling that the reward was not worth the trouble that such a find would bring—either because of the misfortune which, superstition told them, would ensue, or on account of the digging and carrying which they would be called upon to do. This attitude was largely the product of the Adams expedition, during which the Yakuts had been forced to work and provide haulage, which they deeply resented. As a result, many later finds were kept secret. Rather than report it, one police chief ordered all the pieces of a mammoth find to be thrown into the sea. The 1,000 roubles were paid only once, for the Dima carcass (see pp.60–61), together with a silver medal.

In 1910 a new law was passed, designating all mammoth remains as national property and obliging discoverers to report them: but of course it was unenforceable. Posters and leaflets, such as the one above, continued to be printed by the Academy at irregular intervals—in 1924, 1938, and 1973—to encourage people to look out for and report finds. In 1948 a special committee for the study of mammoths and mammoth fauna was created by the USSR Academy of Sciences, based at the Zoological Institute in St Petersburg. Its aim was to act quickly on new information about finds of frozen carcasses. But even today, probably only a small fraction of finds are reported. According to Siberian geologists, work at the gold mines uncovers frozen remains every year, but since the arrival of scientists can delay and complicate the mining, most are lost to science.

In 1983 a group of Siberian construction workers claimed to have fed their dogs with meat from an unearthed frozen mammoth rather than report it to the authorities. Such an attitude is ironic, since a whole preserved mammoth could now fetch a small fortune on the international market.

While chance finds still lead to major discoveries, recent years have seen the growth of dedicated scientific expeditions for the discovery of mammoth and other remains in promising areas of Siberia, allowing the excavation and preservation of carcasses under controlled conditions. Since the 1990s, expeditions to the New Siberian Islands have produced a number of mammoth carcass parts as well as large quantities of skeletal material. Another area subject to intense study has been the Taimyr peninsula of north-central Siberia, where a series of expeditions has recovered large quantities of mammoth bones and some frozen tissue (see main text on opposite page).

*This mammoth ulna* from the Siberian permafrost was cut open to reveal the inner cavity—empty in most fossil bones—still filled with marrow.

encountered them very rarely: Herz reported that one local Lamut, aged over 90, said he had never seen a mammoth before the Beresovka find. When one takes into account the various reasons why finds were never reported, it becomes evident that the frozen carcasses known are merely the "tip of the iceberg." It is likely that many of the innumerable tusks found throughout the centuries were taken from well-preserved carcasses. The very few preserved carcasses we do have are therefore enormously precious, providing science with a unique glimpse of a vanished species.

## THE BABY MAMMOTH DIMA

In June 1977, Alexei Logachev, a bulldozer driver, made one of the most remarkable finds of modern times when he spotted a mammoth calf on Kirgilyakh Creek, a tributary of the Berelekh River on the Upper Kolyma in the Magadan region. He received the 1,000 roubles reward for the discovery. Gold is found at the bottom of valleys, in the sediments of the rivers that flowed here before the Ice Age climate became too cold and dry. To reach them, gold-miners need to remove the solifluction deposits, which they do by hosing the permafrost layers with water jets. It was during such an operation that the mammoth was uncovered.

The complete frozen carcass lay on its left side under 6 ft 6 in (2 m) of frozen silt. Unfortunately, the bulldozer blade cut off part of the right side. This mammoth, named Dima after a local stream, was a male aged about 6–12 months, about 3 ft (90 cm) tall, and 3 ft 8 in (110 cm) long and dates to at least 40,000 years ago. Its emaciated body, including the penis, was well preserved, including all the internal organs, and is the most complete mammoth yet to be recovered. It included parts of the woolly covering on the ears, the 58cm trunk and the body, although only the hair on the feet remained after the subsequent embalming process. Lots of wool was found frozen into the ice below the body. There is a wound on the right foreleg. When alive, the baby mammoth would have weighed 220–250 lb (100–115 kg). However, the dehydrated carcass weighs only 134 lb (61 kg).

His main diet was probably still mother's milk, though the slightly worn milk molars indicate that he had started to feed on vegetation too. The colon contained 7½ lb (3.4 kg) of plant detritus indicating a swampy tundra environment, but there are no plant remains in the small stomach, which contained only

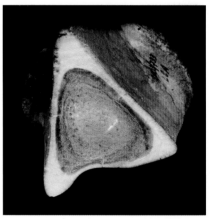

*This mammoth ulna from the Siberian permafrost was cut open to reveal the inner cavity—empty in most fossil bones—still filled with marrow.*

*The tusk of a woolly mammoth discovered during mining operations at Last Chance Creek, near Dawson City in the Yukon Territory of Canada. The 26,000-year-old carcass of a horse, still with food in its stomach, was found nearby.*

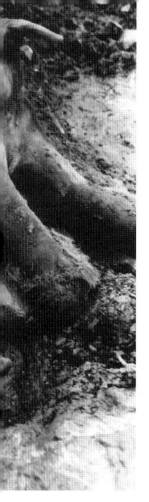

*Like most of the best-preserved frozen carcasses*, the baby male mammoth Dima appears to have died in August, or the early fall. The work teams and bulldozer crews involved in his discovery were specially rewarded for accepting the loss of a few days of gold mining and tolerating the influx of scientists, reporters, and curious onlookers attracted to this remarkable find.

2 oz (50 g) of dark silt together with some of Dima's own hair, and the small intestine is empty. These facts, together with the lack of layers of fat, indicate poor condition. Death was by falling into a pool or crevasse or becoming stuck in a bog or mudflow. The following year Dima was further buried by mud and rock flowing down in the spring thaw.

It has also been suggested that Dima died in a situation from which a normal infant could have escaped. But Dima was sickly, as evidenced by his heavy load of parasites, and lacked the strength to extricate himself. Young animals are frequent victims of such incidents, and the presence of his mother would have kept predators and, after death, scavengers, at bay.

Dima was flown to Leningrad (St. Petersburg), and soon became a worldwide media sensation. Following chemical treatment, it was possible to carry out a full study of his anatomy, including a close inspection of his internal organs, a biochemical examination of the muscles and brain, and a study of his blood vessels and cells (see p.87). Dima has subsequently been exhibited in Europe, North America, and Japan and has been insured for $12 million.

## FINDS FROM ALASKA AND THE YUKON

The explorer Otto von Kotzebue, on a voyage of discovery in 1815–18, wrote about quantities of mammoth teeth and bones exposed by melting ice in Alaska. Since then there have been countless finds of bones, teeth, and tusks, but very few flesh remains, although these appear occasionally.

As in Siberia, it was the start of gold-mining that sharply increased the number of discoveries, since the diggers have to thaw and wash away layers of soil with high-pressure jets to reach the gold-bearing levels, and their work extends over large areas. In places there are huge hydraulic operations, which strip off the sun-softened "muck," while drag lines scrape away the underlying gravel down to bedrock.

Alaska's most famous find was made in 1948 at Fairbanks Creek, where the face, trunk, and one foreleg of a mammoth calf were recovered. The skin was almost hairless; it was embalmed and sent to the American Museum of Natural History in New York, where it was preserved and displayed (see p.35). In life this baby, nicknamed Effie, would have weighed around 220 lb (100 kg). It has been dated to about 25,500 years ago.

## MAMMOTH "CEMETERIES"

The idea of an "elephants' graveyard"—where elephants choose to go to die—is a myth. But it is a myth that contains a grain of truth, for in Africa and Asia there are many distinct places, mostly dried-out waterholes or riverbeds, where accumulations of elephant bones are found. The places mark the spot where elephants have died natural deaths, either singly or in groups, through drought or some catastrophe such as a flash flood.

## "The bones of at least 156 mammoths lay in the banks of the Berelekh River"

There are, likewise, a number of places, especially in Siberia, where large numbers of mammoth bones have accumulated, but the explanation for this is probably somewhat different. Over thousands of years individual mammoths died while walking along river valleys, or by falling through ice, and their bones were eventually brought together by flowing waters in gullies, backwaters, and oxbow lakes.

The greatest accumulation known is on the Berelekh River, a tributary of the Indigirka. The Berelekh meanders between gently sloping hummocks of permafrost that are constantly being eroded by the river and by thawing. Mammoth bones have been washed out of these hummocks for centuries and redeposited on the riverbed. Between 1970 and 1980 a total of 8,830 mammoth bones from a minimum of 156 individuals was collected by expeditions to the site, including a whole frozen back leg of a mammoth.

*Thousands of mammoth bones have accumulated at the Berelekh "cemetery." Expeditions use hoses to free the bones from the frozen soil. To add to the physical hardships of this task, the area swarms with large and voracious mosquitoes.*

A tusk from the base of the bone layer gave a date of 14,100 years ago, while scraps of skin and ligament from another spot were dated to 16,300 years ago. These results provide some idea of the possible time span represented by this accumulation. The Berelekh "cemetery," therefore, is probably the product of numerous individual accidents that occurred over many centuries—the result of animals becoming stuck in fluid mud or falling through thin ice and drowning (see pp.169–70).

A Paleolithic site, dating to about 12,800 years ago, has been found 390 ft (120 m) downstream, containing stone tools as well as some tusk fragments clearly taken from the accumulation of bones. One shaft of tusk, 3 ft 1 in (94 cm) long, bore a remarkable engraving of a mammoth with extremely long legs (see p.118). The occupants might have been able to use some of the fresher meat from the mammoth site, and certainly used bones and tusks as tools, but it is unlikely that they

*Several of the Sevsk skeletons were almost complete (left), with a high proportion of young animals (45 percent were under 10 years of age). At the Berelekh "cemetery" by contrast, the bones were not found as skeletons, and only 30 percent were less than 10 years old, corresponding to a different mode of accumulation.*

played any major role in the accumulation of mammoth remains, especially since the biggest concentration of bones seems to pre-date this site by two millennia.

In the summer of 1988 a different kind of mammoth "cemetery" was discovered on the left bank of the Seva River in southeastern Bryansk Province, south of Moscow, when a bulldozer excavating a sand quarry picked up several bones. The following year a compact bone layer was encountered, containing between 10 and 15 mammoths. Careful excavation at the "Sevsk cemetery" resulted in the recovery of about 4,000 fragments from at least 33 individuals.

This area used to be a marshy river floodplain. Its rich food resources probably attracted the animals, which may have been trapped by flooding. The carcasses seem to have been brought here during a period of high water and quickly buried under alluvial deposits. Although far fewer bones were found here than at Berelekh, this is the largest natural mammoth accumulation known in Europe. The site's importance, however, lies not in the number of animals but in the fact that much of the material comprises large

*The skeleton of a baby mammoth* excavated in a block at the Sevsk mammoth site near Moscow.

articulated parts of skeletons. Seven almost intact skeletons of all ages, including three babies, were recovered. These may all have been members of a single herd, and even of a single family. In total, the number of young animals at the site (45%) is higher than in most fossil assemblages, but is typical of a wild herd of elephants, again suggesting the sudden death of an entire herd.

The Sevsk mammoths seem to have died within a brief period, and were instantly covered, in full or in part, by the river sand. Their remains have been dated to 16,600 years ago.

In the New World, the Waco mammoth site in central Texas has yielded a remarkable assemblage of 22 Columbian mammoth skeletons. Geological evidence suggests that 15 of the animals, all females and young, were a herd that died in a single event. Remarkably, the positions of the skeletons suggest that the mammoths had arranged themselves in a defensive circle prior to death. A second group of mammoths, six females and a bull, were oriented in a line trailing away from the first group. The sediment around their skeletons suggests they may have been caught in a flash flood that caused slumping of the gully walls as the animals walked along a river channel. Both groups died between 64,000–73,000 years ago, but whether in the same or separate events is uncertain.

*A woolly mammoth skull is removed from silty yedoma deposits in Siberia. The erosion of such deposits can give rise to accumulations of bones mimicking death assemblages or "cemeteries."*

## NATURAL MAMMOTH TRAPS

The most spectacular accumulation of Columbian mammoth fossils was found at a former sinkhole (see box opposite) at Hot Springs, a small town in South Dakota, in the middle of the North American continent. So far, about 55 individual mammoths have been discovered, but this list is growing each year, and it is estimated that at least 100 met their end in the sinkhole.

According to radiocarbon dating, these Columbian mammoths died around 30,000 years ago. It is thought that the trap was operative for about 300 to 700 years before it finally filled up with sediment. The individual mammoths are spread throughout the sediment, indicating that they did not enter at the same time but at intervals. This means that even 100 or so preserved skeletons represent only an occasional fatal incident—the entrapment of one animal every several years on average. Most were young adult males, caught at their most adventurous age. Recent studies of isotopes extracted from annual tusk rings (see Ch. 3) suggest that most deaths were in the spring and summer months, perhaps corresponding to maximum availability of attractive vegetation in the sink hole.

There have been further intriguing finds at Hot Springs. Excavation in 1975 revealed a mammoth molar that appeared different from the others: it looked rather like a tooth of a woolly mammoth, unlike all the other mammoth remains from the site, which were of Columbians. However, the researchers did not dare conclude that woolly mammoths were present on such skimpy evidence. Then, toward the end of the 1987 archaeological season, the partial skull of a woolly mammoth was unearthed, with its teeth in place.

This is one of the very few sites where both species have been found at the same location, but this does not necessarily mean that the two coexisted (see p.35). Both of the woolly mammoth specimens were found rather high in the deposits, perhaps indicating that the woolly mammoths entered the area briefly at a time when the range of the Columbians had receded toward the south.

Similar traps existed wherever mammoths lived. At Condover in Shropshire, England, a woman walking her dog around a gravel pit in 1986 spotted some large bones. The result was the recovery of the skeletons of four or five woolly mammoths, dating to between 12,700 and 12,300 years ago—the first evidence that the species survived in Britain after the retreat of the ice. They lie within a kettle hole, a feature formed as a result of large, buried blocks of ice being left by a retreating glacier. As the block

*The Hot Springs site* has now been roofed over in recognition of its exceptional scientific importance. Some 55 mammoth skeletons have so far been unearthed there. The state of preservation of the bones is outstanding: they include delicate hyoid (tongue) bones, which have survived in place, and even bile stones (like human kidney stones).

# MAMMOTH TRAPS

The excavations at Hot Springs in South Dakota show that the mammoths had become trapped in a geological formation known as a sinkhole—a crater about 130 ft (40 m) long, with steep, slippery sides. The mammoths may have been attracted by the warm waters, or were probing the pond's edge for vegetation or water, then either ventured or slipped into the pool. Once inside, the animals would have been unable to escape and would eventually have died of starvation or drowning.

The skeleton of one mammoth, known as Murray (below), lies sprawled against the side, presumably showing how he collapsed in the effort to struggle free. An artesian spring kept the sinkhole wet; meanwhile, erosion of sediment from the sides covered the mammoth bones for posterity. Kettle holes, formed by the melting of glacier ice, trapped mammoths in a similar manner.

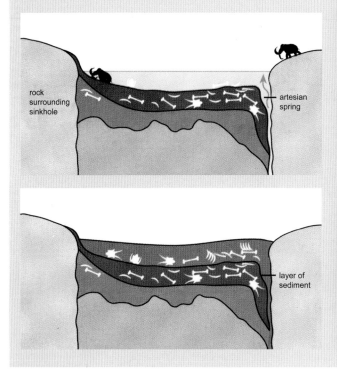

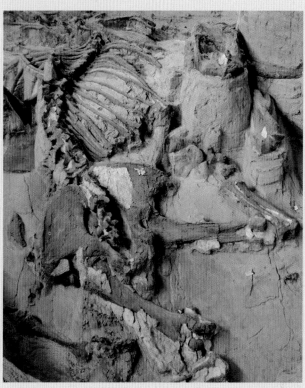

melted, overlying sediments collapsed, forming a crater which subsequently filled up with more sediment. The Condover kettle hole was 33 ft (10 m) deep at its center, but at the end of the Ice Age its sides were not steep. It is therefore likely that the mammoths entered it in search of food or drink and became mired in mud.

The 400 bone fragments found there come from one adult male, aged about 28, and at least three juveniles aged between 3 and 6 years. They were probably not a family group, as juvenile elephants stay with their mother, not their father, while males of this age generally roam alone or in bachelor groups. So the Condover site probably represents occasional deaths over several centuries. The carcasses must have lain exposed, since hatched fly puparia and the remains of dung beetles were found in the bones' cavities.

*Workmen pose with woolly mammoth bones at Condover, England. They had chanced across a complete adult and three juvenile skeletons in an ancient kettle-hole deposit.*

# DEATH IN THE TARPITS

A Columbian mammoth trumpets with fear as it fails to extricate itself from an asphalt seep, while hungry predators look on. This scene is set 30,000 years ago at the tarpits of Rancho La Brea, now in downtown Los Angeles. These natural traps have provided a unique window on late Pleistocene life.

*The La Brea fossils are dominated by the remains of carnivorous species, which came to feed on trapped animals, only to become caught in the tar themselves. A sabre-tooth cat (Smilodon) feeds on a still-exposed mammoth carcass, while another approaches from the left. Dire wolves (Canis dirus) gather excitedly at the scene, while scavenging condors circle overhead or wait in a nearby tree.*

*The accumulated results of 30,000 years of entrapment at La Brea have so far produced some 100 tons of fossils, and more than 600 species of animals and plants. Mammals include wolves, big cats, bison, horses, camels, bears, and the Shasta ground sloth (Nothrotheriops, right), which measured 8 ft (2.4 m) from head to tail.*

*Mammal and bird bones, insect carapaces, and wood* *have been perfectly conserved by asphalt impregnation in the La Brea tarpits. The bones are contained in layers of tar alternating with bands of sand or silt, which were periodically deposited over the tar seeps by streams or winter rains. The remains of herons and pond turtles are testimony to the presence of water bodies nearby.*

## THE LA BREA TARPITS

Mammoths have also been preserved in what is now the heart of Los Angeles, California. Off Wilshire Boulevard are the famous tarpits of Rancho La Brea (*brea* is the Spanish word for tar). These tarpits have been known for centuries and were formerly mined for their natural asphalt. Thousands of tons were extracted before 1875, when it was first noticed that the tar contained fossil remains.

Since then, over 100 tons of fossils, 1.5 million from vertebrates, 2.5 million from invertebrates, have been recovered, often in densely concentrated tangled masses. The creatures found range from insects and birds to giant ground sloths, but a total of 17 proboscideans—including mastodons and Columbian mammoths—have been recovered, most of them from Pit 9, the deepest bone-bearing deposit, which was excavated in 1914. Most of the fossils date to between 40,000 and 11,500 years ago, though the single human find—a woman—dates to about 10,200 years ago.

> ## "Mammoth remains have been discovered in the heart of Los Angeles"

The asphalt at La Brea seeps to the surface, especially in the summer, and forms shallow puddles that would often have been concealed by leaves and dust. Unwary animals would, therefore, become trapped on these thin sheets of liquid asphalt, which are extremely sticky in warm weather. Stuck like flies on flypaper, the unfortunate beasts would die of exhaustion and hunger, or fall prey to predators which often also became stuck.

As the animals decayed, more scavengers would be attracted and caught in their turn. The fact that some bones are heavily weathered and carry fly puparia shows that some corpses remained above the surface for weeks or months. Bacteria in the asphalt would

*Major excavations were undertaken at the Rancho La Brea tarpits in the early decades of the 20th century and established the true significance of this remarkable site. They were found to contain the remains of scores of species of animals from the last 30,000 years of the Ice Age.*

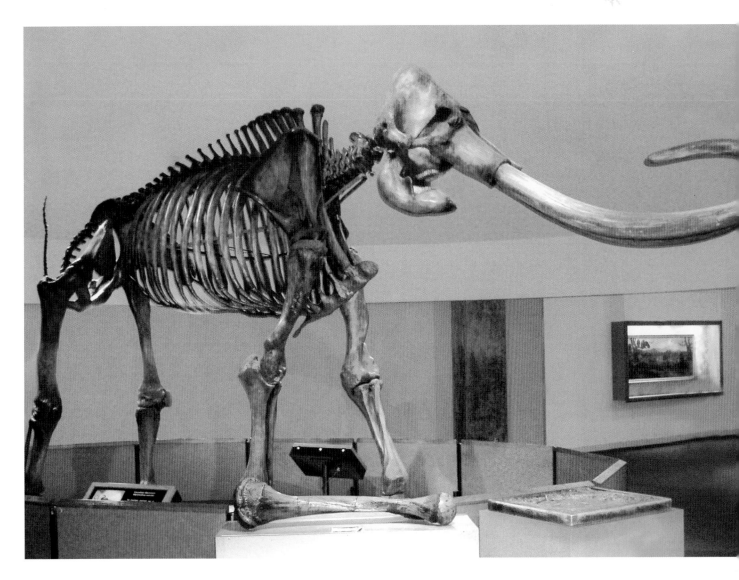

have consumed some of the soft tissues, and the asphalt itself would dissolve what was left, at the same time impregnating and beautifully preserving the saturated bones, which are now dark brown or black and shiny.

No soft tissue survived at La Brea, but it may, in certain circumstances, be preserved by crude oil. At Starunia in Poland the subsoil is filled with veins of paraffin. The site is best known for its preserved rhinoceros carcasses, but in 1907 a mammoth was found at a depth of 140 ft (43 m). There had been some decomposition before the tissues became embalmed, but this was still a mammoth "in the flesh"—the only one outside the permafrost regions—because it had been naturally pickled in a petrochemical seep associated with salt deposits and surrounded by the mineral wax ozocerite. The skeleton was well preserved and the skin still supple, but the hair had become firmly stuck to the surrounding sediments.

*A skeleton of a Columbian mammoth* on display at the Page Museum in Los Angeles.

*The scene at La Brea today* contrasts vividly with that of 90 years ago (top left). Below, life-size figures of Columbian mammoths in and around the tar illustrate one of the many dramas that occurred in this natural trap.

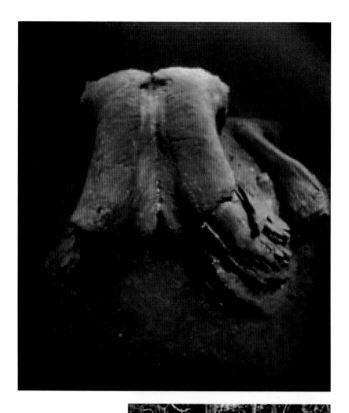

*The skull of a Columbian mammoth*, lying in a bed of silt in the Aucilla River, Florida, was photographed by scuba divers from the underwater archeology team that recovered it.

## RIVER AND LAKE DEPOSITS

While frozen carcasses and natural traps provide the most spectacular source of mammoth remains, the most common and most widespread discoveries are made in a more mundane context: deposits of gravel, sand, or clay being dug up for industrial use or excavated for building foundations or pipelines. These fossils, which fill our museums, are generally isolated teeth, partial tusks, and occasional leg bones. The nature of the sediment gives a clue to the ancient environment of deposition—gravels representing fast-flowing rivers (sometimes bringing meltwater directly from a glacier), sands and silts indicating quieter stretches of rivers, and clays implying still lakes.

> "Countless isolated teeth and bones lie scattered across the whole range where mammoths once lived"

The preservation of isolated bones and teeth in river deposits can be explained as follows. In some cases an animal died by the waterside, and would probably already have been partially scavenged and broken up when, perhaps due to a rise in the river level, it was washed into the water. In other cases the animal may have died while attempting to cross a river, as sometimes happens to elephants in Africa today. In either case, the carcass—or parts of it—would have been further broken up as it was transported downstream before settling, and decay would then have caused additional separation before some elements were finally buried. In lakes, partial or complete carcasses were washed in by rivers, or floated in from a bank if the water level rose. Generally, molars and tusks survive best because their enamel and dentine are harder and denser than bone.

*Five complete mammoth skeletons were retrieved from the Aucilla River, Florida, in the 1960s; they were lifted bone by bone from the riverbed 40 ft (12 m) beneath the surface.*

*The skeleton of a mammoth about 200,000 years old was excavated in 1964 at Aveley on the Lower Thames, east of London. It lay directly above that of a straight-tusked elephant, Palaeoloxodon antiquus, which had occupied the site when it was more wooded earlier in the interglacial period.*

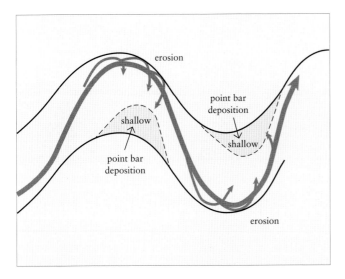

*Bones may be buried* in a river-bottom along the main course of water flow (heavy arrow); or in shallow, quieter bends where "point bars" are being deposited.

Excavations at Stanton Harcourt, near Oxford in central England have provided an unprecedentedly detailed picture of how the animal remains become preserved in river sediments. The deposits, some 200,000 years old, are believed to represent the ancient River Thames. A meandering river carries sediment—silt, sand, or gravel, depending on the size of the river, its speed of flow, and the rock type that it flows over. When in flood it may carry other items such as pieces of wood, and carcasses of animals that have died upstream. Larger, heavier items, such as mammoth limb bones and tusks, would be rolled along the gravel at the bottom of the river. Many of the large bones and tusks at Stanton Harcourt were resting on thin layers of silt, where the current had waned for a while (possibly during summer low water), but coarse sediment became banked up against these large obstacles. The constant movement of sediment up, over and round the bones ensured their rapid burial.

The studies at Stanton Harcourt showed that another kind of "bone trap" occurs along the inner bends of a river, where water flows relatively slowly and sediment may be deposited to form so-called point bars. Large objects such as bones become stranded on the point bars, accumulating in any hollows on its surface, and may then be rapidly covered as more sediment is brought downstream. Eventually, as the river's channel migrates sideways, the point bar is colonized by vegetation. Occasional flood events may still inundate it, incorporating further wood, bones, and other material. The point bar itself can become a focus for animals coming to the water to drink, and any that die there or become mired may contribute their remains to the accumulation. Eventually the river channel leaves the bar as a preserved fossil assemblage.

*The excavations at Stanton Harcourt,* near Oxford, England, have so far uncovered 120 mammoth tusks. During the excavation season of 1993 one of these, encased in plaster, proved so heavy that it had to be lifted from the site by a Royal Air Force helicopter.

The Stanton Harcourt excavations have yielded about 120 mammoth tusks, numerous molar teeth, and rarer remains of other animals such as horse, bison, fish, and frogs. The presence of acorns and hazelnuts indicates that the animals represented in the finds lived during an interglacial period, when the climate was similar to that of today or slightly warmer. The mammoths were a late variety of *Mammuthus trogontherii*, the steppe mammoth, which was not yet fully adapted to cold conditions.

Occasionally, a complete skeleton is found in this type of deposit. Specimens that have been mounted include skeletons from the Aa River in northern France, Steinheim (see p.25)) and Ahlen in Germany, Mátra in Hungary, and Heilongjiang in China. An exceptionally complete skeleton was discovered by a 16-year-old boy while walking in a wood which had grown over former lake sediments, near Siegsdorf in Bavaria. He found the first rib in 1975, but the complete skeleton was not finally excavated until 1985, a task which entailed the removal of 777,000 cubic ft (22,000 m³) of earth and clay deposit by bulldozers. The specimen appears to be the largest known woolly mammoth of the last Ice Age. Now carefully reassembled and mounted, and

*Three tusks formed part of a large assemblage of mammoth remains discovered in river channel deposits at Lynford, Norfolk, England, in 2002.*

christened Oscar, it stands 11 ft (3.4 m) high, with tusks up to 8 ft 6 in (2.6 m) in length. It has been dated by the electron spin resonance technique to around 48,000 years ago.

Finds of otherwise complete mammoth skeletons often lack one crucial element: the skull. This may well be because, full of gases that result from decay, it became detached from the carcass and floated off. (This still occurs with modern elephant corpses in African lakes and rivers.) Sometimes the toughest parts survive, such as the back of the skull, tusks, and teeth, while the thin, porous bones of the rest of the skull have disintegrated. This is the case in several of the skeletons described above, including those from Condover, Steinheim, and Siegsdorf, whose mounted skulls have been reconstructed in plaster or resin. Examples of complete skulls excavated from river deposits are therefore precious, such the exquisitely-preserved cranium of a female woolly mammoth discovered in a gravel pit in Gloucestershire, England, in 2004.

## VOLCANIC DEPOSITS

Perhaps the rarest deposits in which mammoths are preserved are those produced by volcanic activity. At Tocuila, in the basin of Mexico, at least seven mammoths, including massive skulls, were found embedded in a volcanic mudflow derived from the eruption of the Nevada de Toluca volcano, which still towers over the site, around 12,500 years ago.

Below the mammoth horizon are volcanic ashes, deriving directly from the eruption. The mammoth remains are enclosed in a deposit, some 5½ ft (1.7 m) thick, known as lahar. Lahars form when heavy rains or melted snows wash loose volcanic debris down from higher to lower ground. These mudflows are catastrophic events, but the skeletons are disarticulated, suggesting the mudflows did not kill them but merely covered already-dead carcasses. It has been suggested that the mammoths may have died from toxic volcanic gases, and were then picked up by the mudflows and deposited at the site.

*Mammoth skulls* embedded in volcanic deposits at Tocuila in the Basin of Mexico.

## CAVES, DUNG, AND FOOTPRINTS

Mammoth remains are preserved in caves for several reasons but were usually carried in by humans or predators. Juvenile mammoth remains were accumulated in Friesenhahn Cave in Texas, for example, by the extinct scimitar cat, and in Kent's Cavern, England, by the spotted hyena (see p.105). Many of the mammoth bones bear the tooth marks of these predators or scavengers.

Occasionally, mammoths entered caves to escape bad weather, or in search of water or salt. Similarly, modern elephants have been seen going into Kitum Cave, Kenya, in order to scrape salt from its walls with their tusks. A striking indicator of this are the quantities of mammoth dung found in dry caves in the American Southwest, which have provided detailed evidence of diet (see p.90). In the Chinese province of Inner Mongolia, 7 lb (3 kg) of dung was recovered in 1980 from an open-cut coal mine, together with bones of two mammoths.

Another rare but remarkable trace of mammoth presence comes in the form of preserved footprints. The most remarkable were discovered in 1999, when the water level of St. Mary's Reservoir, southern

*Bones recovered from the North Sea have often been blackened by mineralization, but to preserve them they merely require desalinating by soaking in fresh water. An average "catch" from a six-week trawl (above) includes bones of reindeer and bison as well as mammoths.*

*A fine catch. This skull of a woolly mammoth was dredged by a fishing boat from the floor of the North Sea between Britain and the Netherlands.*

Alberta, was lowered. This exposed areas of the reservoir floor to wind erosion which removed 4 ft 9 in–6 ft 6 in (1.5–2m) of sediment and exposed tracks, fossil bones, and archaeological material. The tracks include those of camels, bovids, horses, deer, and mammoths. The shallow footprints had been made in a muddy surface of wet, wind-blown silts. There are approximately 500 footprints, many of them organized into trackways. The large, circular prints, with three or more toe impressions at the front, are unquestionably those of a proboscidean, and woolly mammoths are the most likely culprit as they are commonest fossil proboscideans in the region. Their age is around 13,000 years, and they allow some fascinating deductions of mammoth behavior (see Ch. 3).

At the Hot Springs Mammoth Site, footprints can be seen in vertical sections of the excavation. Very different from the Canadian situation, these are individual deep, prints, the result of a Columbian mammoth attempting to extricate itself from the soft, saturated mud that claimed many individuals.

## FISHING FOR MAMMOTHS

Perhaps the most unexpected source of mammoth bones is the bed of the North Sea. At times during the Ice Age, when the sea level was 330 ft (100 m) lower than it is today, the North Sea was an area of dry land. During those periods, animals would have died there, and been deposited in rivers and lakes in the usual way.

Most of the remains are from the typical fauna of the last Ice Age in Europe (110,000 to 11,500 years ago), and include bison, horses, woolly rhinos, reindeer, and giant deer; but proboscideans are common, especially woolly mammoths. A few remains of the ancestral mammoth, and even a form of mastodont, both over a million years old, have also been recovered. The bones are all separate, but are well preserved.

Nets designed to catch flatfish (sole, turbot, plaice) trawl across the uneven seabed, and the beams at the front rake up the fish so that they pass back into the net. At the same time they dredge up—and sometimes break—the fossils that have been laid bare by currents. The bones come from depths of 65–165 ft (20–50 m). Each haul takes about an hour and covers 6 miles (10 km). A single fishing trip can produce several hundred fossils, and every year a fishing boat has been hired by Leiden's Natural History Museum specifically to bring up a catch of fossils. Amateur fossil hunters haunt the ports, awaiting the return of the boats with their bones, which make an odd sight when encrusted with sea barnacles and worm tubes.

Closer to shore, a more recently-discovered fossil source lies five miles west of Rotterdam Harbor, on the bottom of the so-called "Eurogeul" (dredged shipping lane). The geology consists of submerged river sands, and numerous mammal bones, including those of woolly mammoth, have been "fished up" by getting entangled in nets. The age of the material is in excess of 40,000 years.

Large collections of fossil mammal bones have also been dredged from the Penghu Channel, between Taiwan and mainland China. These include a few complete molar teeth of mammoth, identified for the first time in 1999.

## THE WEST RUNTON DISCOVERY

On December 13 1990, two local naturalists were walking along the beach at West Runton, Norfolk, England, when they spotted a dark bone protruding from the cliff. When exposed, the fossil proved to be one of the largest of mammalian bones: the complete pelvic girdle of an adult male steppe mammoth. Over the following years, further mammoth bones were discovered in the same spot, leading to the realization that a complete skeleton might be preserved in the cliff, and in 1995 a major excavation was organised at the site.

The deposits at West Runton, famous for their fossil remains, were laid down during a warm interglacial period some 700–600,000 years ago. The skeleton, when uncovered, was partially scattered, but it was clear that the mammoth had died slumped onto its front, in shallow water at the edge of a slow-moving river. Chewed bones and perfectly-preserved fossil droppings indicated that hyaenas had fed on the carcass before burial (see p.105). Some of the bones had been crushed and pressed into the soft sediment, presumably by the trampling activities of other mammoths.

As excavation proceeded, the position of every bone was carefully recorded before the specimen was encased in a plaster jacket and removed from the site. Samples of sediment were collected for extraction of insect and plant remains, later to provide a detailed picture of the mammoth's environment. Back in the laboratory, the bones were painstakingly conserved and prepared for scientific study. The form of the jaw bone and teeth (see pp.26–27) confirmed the 13 ft- (4 m-) high skeleton as that of a steppe mammoth, *Mammuthus trogontherii*, one of only two or three complete specimens known, and thus of great scientific importance in understanding the origin of its direct descendant, the much more familiar woolly mammoth *M. primigenius*.

A further surprise was in store when the hind leg bones were exposed. The right knee joint was extremely abnormal, the result of injury and infection. Detailed examination by veterinary specialists suggests that, debilitated by its injury, the animal slipped, fell and, being unable to get up again, expired by the river (see Ch. 3).

*The excavation of the West Runton mammoth in 1995. The 4 ft 9 in- (1.5 m-) high skull, with its massive 6 ft 6 in- (2 m-) long tusk in place, sits an upright position. The other tusk was detached and crushed, perhaps trampled by other mammoths.*

*Aerial view of the excavation. Thousands of tons of overlying cliff had to be removed before exposing the mammoth skeleton within the rectangle of dark sediment at the foot of the cliff.*

# THE NATURAL HISTORY OF MAMMOTHS

More is known about the appearance and way of life of the mammoths than about any other extinct prehistoric animal. Almost all other fossil species are represented only by their bones and teeth. But for the mammoth—thanks to the preservation of remains frozen in the far north and dried in caves farther south—there is flesh, hair, stomach contents, and even dung. In addition, our human ancestors left us a record of the mammoths they saw, in cave paintings and other art forms. By combining these sources of information with deductions based on what is known about modern elephants, it is possible to arrive at a genuine "natural history" of the mammoths that is more complete than those we have for many animal species still alive today.

*Prehistoric naturalists have bequeathed us this mammoth drawing* at Rouffignac Cave in France. It shows an encounter between two mammoths, similar to the greeting ceremony commonly seen in female elephants, or maybe a pushing contest between two males. Such evidence contributes to our perception of the mammoth as a living animal.

# FUELING THE BULK

Eating took up a great deal of the mammoth's life. The large Columbian mammoth, shown here, needed some 500 lb (225 kg) of fresh food a day, and so was almost continually foraging. This scene is set about 18,000 years ago on the Colorado Plateau in the southwestern United States, where the mammoths' feeding habits to some degree played a part in shaping the local vegetation.

*After eating the bark* which it stripped with with its tusks, a mammoth pushes over a small tree to gain access to the upper foliage. The tusks could also be used for digging up plants by the roots.

*A female pulls up a bunch of grass* by curling her trunk around the stems, while a youngster pulls a few stalks from her mouth. Two mammoths are drinking from a stream, sucking up water with their trunks and then squirting it back into their mouths.

THE NATURAL HISTORY OF MAMMOTHS

*With its thousands of small muscles*, the trunk was capable of delicate operations. The two long projections at the tip acted like a "finger and thumb" to pick flowers, buds, or grass stems. Possibly these evolved as an adaptation to feeding on the short grasses of the Ice Age, as compared to the tall tropical grasses eaten by modern elephants, which use mainly the "trunk curling" method.

*All mammoths were completely vegetarian.* The precise composition of their plant food varied with the local conditions, since both the Columbian and the woolly mammoth occupied a wide range of habitats. However, grass was almost always the staple, supplemented by flowering herbs, shrubs, and parts of trees as available.

*Many other herbivorous species* shared the local forage: above, a group of now extinct American camels grazes on the other side of the stream.

THE NATURAL HISTORY OF MAMMOTHS 79

# NOT WITHOUT ENEMIES

Defending themselves from predators was a relatively easy task for mammoths, with their powerful trunk, deadly tusks, and huge body. Nonetheless, a group of mammoths needed to remain vigilant, since pack-hunting carnivores could occasionally pick off a youngster or ailing adult.

*In this scene, set in Alaska during the last Ice Age, a pack of wolves harries a small group of woolly mammoths. As relatively small predators, wolves would rarely have troubled mammoths, and certainly could never have taken an adult in its prime. This pack, however, by distracting and confusing a small mammoth group, may have been attempting to capture—most likely without success—the baby being held back and protected by its mother.*

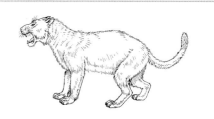

*Large carnivores such as lions* and hyenas are today restricted to tropical areas. This is not because of climate, but because only in the tropics does abundant large herbivore prey remain. In the Pleistocene many more herbivorous mammals still lived in the northern continents and were preyed upon by hyenas and various large cats which are now extinct, including the lion Panthera leo atrox *(right) and the sabre-tooth and scimitar-tooth cats.*

*With a body weight measured in tons*, mammoths were extremely strong animals, and a healthy adult could deal with a challenge from almost any other creature. A sideways or downward swipe with the tusks would stun or kill an aggressor and, by lowering the head, the points of the tusks could be used for stabbing. The trunk, as in living elephants, was extremely powerful and could kill a predator by breaking its back. Finally, an attacker would have to avoid being jumped on or trodden on.

THE NATURAL HISTORY OF MAMMOTHS 81

The most striking feature of the mammoths was their sheer scale. They were big animals, the biggest in their environment, and quite accurate estimates of their size can be obtained from their mounted skeletons, with a little added for flesh (see p.174). That said, most mammoths were not larger than living elephants. In the woolly mammoth, adult males generally stood between about 9 and 11 ft (2.7 and 3.4 m) at the shoulder. This size compares quite closely with the living African elephant, the world's largest land mammal, in which heights of 10–11 ft (3–3.4 m) are common in males. As is the case with living elephants, female mammoth skeletons are noticeably smaller and more lightly built than those of males. Woolly mammoth females probably averaged 8 ft 6 in–9 ft 6 in (2.6–2.9 m) in height. Using elephants as a guide, a body weight of up to 6 tons in male woolly mammoths can be calculated—about 80 times the weight of an adult human of 170 lb (77 kg).

## "The Columbian mammoth weighed the equivalent of 130 adult humans"

Despite its great size, the woolly mammoth, *Mammuthus primigenius*, was in fact smaller than its immediate ancestors, and so does not represent the continuation of an evolutionary trend toward greater size. Skeletons of the ancestral mammoth *M. meridionalis* and steppe mammoth *M. trogontherii* indicate an animal that was generally about 13 ft (4 m) high. The reasons for the reduction in size to *M. primigenius* are unknown. In the American lineage

*Baby mammoth Dima shows several characteristic woolly mammoth features, such as the very short tail. The X-ray of his skeleton reveals how the domed skull and shoulder hump had not yet formed, and shows limb bones with free ends, allowing space for growth in length.*

this was less evident: *M. columbi* had a shoulder height of up to 13 ft (4 m), and a weight of about 10 tons, equivalent to around 130 average adult humans.

The outline of the mammoth's body was distinctive: it is shown consistently enough in dozens of Ice Age depictions (see Ch.4) to indicate that the shape given was genuine and not the result of an individual artist's imagination. The depictions present a relatively large head, with a high dome on top. The neck, although short, forms a very clear division between the head and the back. The shoulder has a distinctive, high hump, behind which the back slopes markedly from front to rear. Preserved skeletons show that the sloping back was due mostly to a decrease in the length of spinal vertebrae, rather than to any difference in the length of front and back legs.

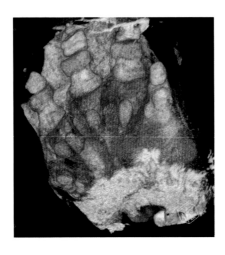

*A CT scan of the Yukagir mammoth's foot (left), showing toe bones. The mammoth walked on tip-toe, supported by a large, springy pad behind the bones.*

Like elephants, but unlike most other mammals, mammoths continued to grow well into adult life. Skeletons and frozen carcasses of young animals give clues to the mammoth's growth (the age of these individuals can be determined from their teeth, see p.93). Most individuals probably weighed about 200 lb (90 kg) at birth. The famed frozen mammoth Dima (see pp.60–61) was 6 to 12 months old at death and was about 3 ft (90 cm) high. He shows that the shoulder hump and sloping back developed later in life: in infants the back formed a convex arch rather like a modern adult Asian elephant's. By the age of ten, a mammoth's height had doubled and its weight increased 15-fold. Studies of bone development indicate that the ends of some limb bones did not fuse to the main shaft until about 40 years old in males and 25 in females, so the limbs were still lengthening up to that age. Life expectancy was about 60 years for the woolly mammoth, but was probably higher for the larger Columbian mammoth.

The exquisite preservation of the Yukagir mammoth shows adaptations to locomotion similar to those of living elephants. The sole of the foot is dissected by numerous cracks producing a "crazy paving" effect,

*The foot of the Yukagir mammoth* is excavated in 2003, revealing the exquisite preservation of its sole, similar to that of living elephants.

*The coarse, wiry outer hairs* of a woolly mammoth (shown actual size). The orange color is probably not natural, but the result of long burial.

and helping the animal grip the surface while walking or climbing. CT scans of the animal's foot further reveal the presence of a large pad of tissue behind the toes, which compressed when the mammoth put its weight on the foot, and re-expanded on lifting it, giving the animal a "spring in its step." Several of the images of mammoths from Paleolithic art show the foot swollen, as the foot pad compressed under the animal's weight. Examples are those from Cussac, Rouffignac, and more schematically, the "moon-buggy" mammoth from Chauvet (see pp.91 and pp.120–1).

Many features of the woolly mammoth show striking adaptations to an Arctic climate. There is much less information available for its American contemporary, the Columbian mammoth, and for the earlier species in Eurasia, since soft tissues are not preserved, but it is likely, given their warmer habitat, that these adaptations were less developed.

The furry coat of the woolly mammoth is known from many Siberian carcasses. The most common color is orange, but it varies on different specimens from blond to brown or almost black. Although many modern reconstructions show the mammoth with an orange coat, it is likely that the predominance of this

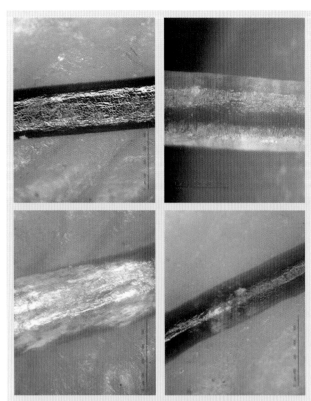

Mammoth guard hairs magnified 30–50X under the microscope show a surprising range of color, from almost black, through brown and orange to blond. There was variation even within single strands.

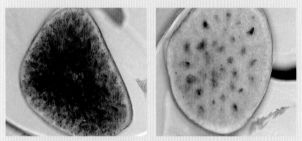

Cross-sections of mammoth guard hairs, magnified about 100X, show how dark hair results from dense packing of pigment grains throughout the strand, while pale hair has many fewer, dispersed, pigment grains.

The tangle of thin strands forming the mammoth's underwool, magnified 75X.

color in preserved hair is largely the result of loss of natural pigment during long burial. Dark brown seems the most likely general color, although no doubt there was variation.

The mammoth's fur consisted of long, coarse outer guard hairs beneath which were shorter, thinner hairs forming an underwool. The guard hairs were typically about one hundredth to one-fiftieth of an inch (0.25–0.5 mm) in diameter, about three to six times thicker than a typical human hair and one or two times as thick as those of an Asian elephant. In their preserved state, at least, they are springy and transparent, resembling fishing line. The hairs of the underwool were much thinner and shorter—about 1–3 in (2.5–8 cm) long—but they were more densely packed and formed an effective insulating layer.

In 2003, the microscopic structure of mammoth hair became a subject of study for "hair scientists" at a leading cosmetics corporation. They discovered that coloration of a single hair is not constant: pigmented brown layers, about 4 in (10 cm) long, alternate with paler zones, beige to yellow, ¾–1½ in (2–4cm) long. The pale zones are more fragile than the darker ones. The fine under-fur, only 1/500 in (0.05 mm) in diameter, is gently curly.

All parts of the body were covered with fur, ranging in length from a few inches to over 3 ft (90 cm). The head, including ears and trunk, was clothed in relatively short hair a few inches long, although it was longer under the chin and neck and on the sides of the trunk, from which hung a clear fringe, noted by Ice Age artists. Hair on the upper body was 1 ft (30 cm) or so long, but from the flanks and belly hung the longest hair, up to 3 ft (90 cm) long, forming a "skirt" reminiscent of the living musk ox or the Tibetan yak.

The upper leg bore fur up to 15 in (38 cm) in length, and several frozen carcasses show that even on the feet the hair was 6 in (15 cm) long, falling to the ends of the toes. It is likely that the woolly mammoth changed coats between winter and summer, shedding its heaviest fur in spring. Since most of the frozen carcasses are probably of animals that died in the fall—when the soil was still mobile but the falling temperature allowed fast freezing—much of the evidence of the mammoth's fur may be based on its winter coat.

Little evidence is available on the coat of the Columbian mammoth of North America. However, in

Bechan Cave in Utah, noted for its impressive deposit of mammoth dung (see pp.89–90), quantities of hair were discovered, including coarse material identical to that of woolly mammoths from Siberia. Since Bechan Cave is in the southwestern United States, the hair almost certainly pertains to *M. columbi*, and indicates that that species was not naked, although the density and distribution of its fur is unknown.

The mammoth's skin was between ½ and 1 in (1.25 and 2.5 cm) thick, no different from that of living elephants. In the case of the woolly mammoth, a thick fat layer 3–4 in (8–10 cm) deep lay beneath the skin, as seen in the carcasses from Beresovka and the Liakhov Islands. This would have contributed to heat insulation, although fat layers are known to be less effective in this function than a hairy coat.

In addition to the fur and fat layer, there were several other features of the woolly mammoth that were probably adaptations to a cold climate. Most obvious among these are the small size of the ears and tail. In living elephants the large ears act as a radiator and are flapped in hot weather to help the animal lose heat.

*Living elephants have hair over much of their body, most notably on the head and hanging from the end of the tail. The potential of elephants and their relatives to produce a furry coat, shown clearly in this baby Asian elephant, was developed to the greatest degree in the woolly mammoth, as an adaptation to much colder conditions than the tropical habitat of the modern species.*

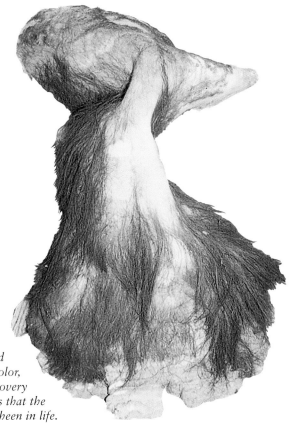

*A frozen carcass of a woolly mammoth, discovered in the Liakhov Islands off northern Siberia in 1901, included a complete foot. Although twisted and dehydrated, it still shows its horny nails, dark coat color, and long hair reaching down to the toes. The discovery of sebaceous (oil) glands in mammoth skin indicates that the coat would have had a glossy sheen in life.*

The woolly mammoth, living in a cold climate, had the opposite problem, and the small size of its ears and tail helped to prevent the loss of body heat. Even more significant, there was a danger of frostbite on these thin, exposed organs, which their small size and hairy covering would have minimized.

The woolly mammoth's ears, preserved on at least four known adult carcasses, were surprisingly similar in shape to human ears. They were about 15 in (38 cm) long and 7–11 in (18–28 cm) across, with a total area of only one-fifteenth that of an African elephant (see opposite and p.103). In the case of baby Dima, the ear was less than 5 in (13 cm) long. In living elephants the ears are used not only for hearing and heat loss, but also to intimidate rivals or predators, the ears being held out from the body during a mock charge. Given the reduced size, this function was presumably less significant in the woolly mammoth.

Until recently, a complete tail was preserved only in the Beresovka carcass and baby Dima. For Otto Herz, leader of the Beresovka expedition (see pp.52–53), the discovery of the tail was a moment of triumph. He measured its exposed length as just 14 in (36 cm). The Beresovka skeleton shows that this relatively short length results from the fact that the tail contains only 21 vertebrae, compared to 28 to 33 in modern elephants. The short, stubby tail of the mammoth was confirmed by the 2003 discovery of a perfectly-preserved specimen on Bolshoi Lyakhowski Island in the Arctic Ocean. Several Ice Age depictions illustrate this feature, showing the fleshy part of the tail extending only a short distance down the leg, in contrast to living elephants where it hangs well below the "knee."

Several carcasses also show, however, that the short, fleshy part of the tail was extended by long, coarse hairs that were up to 2 ft (60 cm) in length and twice the thickness of normal guard hairs. These hairs, hanging in bunches from the tail, would have compensated for the shortness of the fleshy part of the tail and ensured that the mammoth retained an effective fly-swat. Again, this feature can be perceived in Ice Age art, especially the engraving from La Madeleine (see p.117) and drawings from Rouffignac and Cussac (see p.121).

In the Columbian mammoth, the tail seems to have been less reduced than in its woolly cousin. Tail vertebrae are rarely all preserved in excavated skeletons, but in the Huntingdon mammoth, Utah, a series of nine, from the mid-part of the tail, were found. They measure 2½ ft (76.5cm), implying that the tail of the Columbian mammoth was intermediate in length between those of the woolly mammoth and living elephants, in keeping with its less extreme cold adaptation.

At the base of the woolly mammoth's tail was a broad flap of skin covering the anus. This has sometimes been regarded as a further adaptation to the cold, but it is also present in living elephants that inhabit a tropical climate. The anal flap of the mammoth was faithfully recorded by several prehistoric artists.

*A local Siberian poses beside a section of the Beresovka mammoth, found in 1901. Its short tail, at the top of the picture, has at its base the anal flap. The 3-ft (86-cm) long penis projects toward the bottom left.*

THE NATURAL HISTORY OF MAMMOTHS

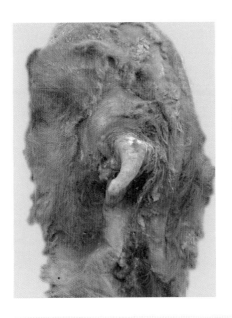

*A hundred years after the discovery of the Beresovka tail, this perfectly preserved (far left) specimen was found on Bolshoi-Lyakhowski Island in the Arctic Ocean. Left, the head of the Adams frozen mammoth, exacavated in 1806, still retains some of its dried flesh and skin. The top left corner of the picture shows the perfectly preserved, oval-shaped ear, only 15 in (38 cm) long. The tusks emerge from their sockets at bottom right. Another perfectly-preserved, and very small, ear is seen on the Yukagir carcass (p.103).*

# UNDER THE MICROSCOPE

The mummification of carcasses in the permafrost was so effective that not only organs but also tissues and cells are extraordinarily well preserved.

The very latest non-destructive scientific techniques have been applied to the most recent finds of intact mammoths. For example, Japanese investigators have produced a three-dimensional picture of the skulls of Mascha (see p.58) and the Yukagir mammoth (see p.103) by computer tomography (CT). In this method, the scanner provides color images of cross-sectional "slices" through the body, allowing detailed internal views without opening it. Dima's heart is still in excellent condition, and the Japanese anatomists have likewise been able to produce colored images of "slices" of the organ, as well as a three-dimensional computer image through tomography. Despite some shriveling and wrinkling of the tissues, all of the heart's essential internal structure could still be seen.

Dima's preserved carcass also provided scientists with a unique chance to investigate such details as blood cells after more than 40,000 years of burial in the permafrost. Under a scanning electron microscope, red and white cells are indistinguishable from those of living

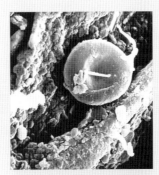

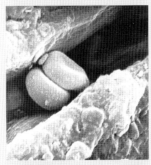

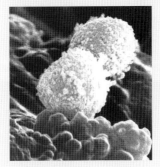

mammals. Microscopic studies also revealed parasitic flies and protozoa in Dima's gut. In 1976 two U.S. scientists reported on the tissues of the mammoth face from Alaska dating to about 25,500 years ago (see p.35). Although the skin was dry and leathery, the hair was well preserved, and the eyes survived as globoid structures filled with a soft, white, cheesy material.

In 2004, thin sections of the skin of a Siberian carcass were examined microscopically. The study revealed, in the deeper layers of the skin, sebaceous glands, sweat glands, hair follicles, and small blood vessels. This finally resolved a long-standing debate as to whether mammoths possessed sebaceous glands. These glands produce a glossy, water-repellent covering to hair, an important adaptation to withstanding cold, wet weather.

*Dima's blood cells, magnified about 2,000 times by electron microscope, show two white cells (bottom); two red cells in a blood vessel (middle); and a red cell with the characteristic ring shape and a fungal tube growing into it (top).*

## HABITAT AND DIET

The main habitat of the woolly mammoth during the last Ice Age, between about 100,000 and 14,000 years ago, was the vast expanse of grassy vegetation that covered much of Europe, northern Asia and northern North America. This vegetation has no precise modern equivalent. It is known as "mammoth steppe," or "tundra steppe," because it resembled in some ways the grassy steppe of southern Russia today, although the vegetation was more diverse than in this region. Contrary to popular belief, the mammoth did not live in a habitat dominated by ice and snow. The continental landmasses were expanded by lowered sea levels, and the weather is believed to have been dominated by high pressure systems, so little rain or snowfall reached their interiors.

> ## "The stomachs of several frozen mammoths contain the remains of their last meal"

The plants of the dry grassland of the "mammoth steppe" were faster growing and more abundant than today's tundra plants. In addition to various species of grasses, they included dry-ground sedges, small shrubs, such as Arctic sagebrush, and many herbaceous plants including members of the pea, daisy, and buttercup families. This was the habitat that supported large numbers of woolly mammoths, as well as other grazers such as bison, horses, and woolly rhinoceroses. Large areas of the landscape were treeless, but scattered birch, larch, and other trees occurred locally, especially in more southerly regions.

Details of the "mammoth steppe" vegetation have come from plant remains preserved in dated fossil deposits (see pp.172–3). However, we also have a more direct and dramatic source of evidence about the mammoth's diet. Several of the frozen carcasses found in Siberia include the animal's stomach and intestines containing the remains of its last meal. Most spectacular among these is the Shandrin mammoth, whose entrails contained no less than 640 lb (290 kg) of partially digested food, resembling densely compressed hay. The Yuribei mammoth, a relatively late carcass dated to around 11,500 years ago, still had green grass in its stomach. When the Beresovka mammoth was discovered, it even had food between its teeth and on its tongue, the squashed grasses still bearing the imprint of the animal's molars.

Detailed botanical investigation of these remains has provided important clues to the mammoth's diet. Leaves, seeds, fruits, and pollen can be identified to at least the general group of plants (e.g. grasses, buttercups), and sometimes to the exact species.

These analyses show that the woolly mammoth's diet was dominated by grasses and sedges. For example, the food in the stomach of the Beresovka mammoth comprised largely grass, with additional remains of various herbs. The guts of the Shandrin mammoth contained 90 percent grass by volume, with a few twig tips of willow, birch, larch, and alder. Studies on the stomach contents of the Yukagir mammoth (see pp.56–58), announced in 2005, show a dominance of grasses, supplemented by an array of steppic plants and shrubs, but in this case, absence of trees. Willow twigs formed an important component of the meal, but this was a small-leaved, dwarf species. The mosses indicated a variety of environments, ranging from moist to dry conditions. Mammoths clearly liked to supplement their grassy food with herbaceous plants and shrubs, and even the occasional tree-browse when available. Like living elephants, they needed this variety to provide the different nutrients necessary for growth, and the details of these supplements probably varied considerably, depending on the local vegetation. The stomach

*A slice through the intestines of the Shandrin frozen mammoth, unearthed in Siberia in 1972, shows how the animal's gut was completely packed with chewed-up, fibrous vegetable food. Detailed study of the leaves, seeds, and pollen has revealed much about the woolly mammoth's diet.*

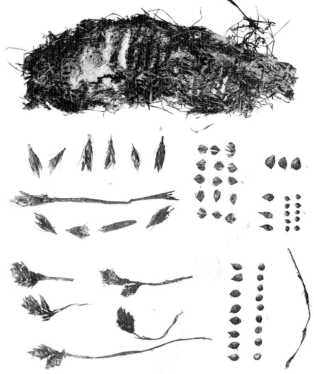

contents of the Yukagir mammoth contained another surprise: spores of dung-inhabiting kinds of fungus. The spores had been ingested together with the food plants, their presence reflecting the availability of herbivore dung as a substrate for their own growth.

The Columbian mammoth, inhabiting the southern half of North America, probably covered a greater variety of habitats in the last Ice Age than did the woolly mammoth farther north. Many regions are thought to have been clothed by a "mosaic" vegetation of grasses, herbaceous plants, shrubs, and trees. In some areas this would have been predominantly grassy and meadowlike, with trees and shrubs concentrated along river courses; in others the trees and shrubs were scattered more evenly, forming a savanna-like or "parkland" landscape. Elsewhere there were coniferous or deciduous woodlands with extensive understorey. Only areas of true forest would not have been favored by the mammoths, although even here locally open areas could have provided grass and herb graze.

Since no carcasses of Columbian mammoths have been preserved, stomach contents are very rare. When the Huntingdon mammoth, Utah, was excavated, pieces of compact vegetation, comprising masticated organic material, were found in the sediment between the pelvis and ribs. Microscopic analysis of the plant remains revealed that the mammoth had eaten sedge, grass, fir twigs, and needles, and even a little oak and maple.

An even more startling source of information about Columbian mammoth diet has come from two caves in southern Utah. These have yielded large quantities of mammoth dung, giving a direct insight into the diet of *Mammuthus columbi*. The dry conditions and the relatively uniform temperature in these caves on the Colorado Plateau have allowed the dung to survive without decay for thousands of years. At Bechan Cave a dung blanket 16 in (41 cm) thick, with a total volume of 8,000 cubic ft (227 m³), has been excavated.

The Bechan deposit includes droppings of various smaller mammals, but the most spectacular are large, spherical boluses about 8 in (20 cm) in diameter. These are very similar to modern elephant dung and so are identified as belonging to the Columbian mammoth, the only possible candidate among the local Pleistocene fauna. The quantity of dung in the cave seems extraordinary, but since an adult African elephant drops an average of 25 lb (11 kg) every 2 hours, and 200–300 lb (90–135 kg) every day, a small group of mammoths could easily have produced the Bechan deposit in a short time, or over a few seasons. The name "Bechan" is derived from a Navajo word meaning "large faeces."

Bechan mammoth dung has been dated to between 16,000 and 13,500 years ago, and comprises 95 percent grass and sedge by weight. Woody plants also occurred, in quantities varying from zero to 25 percent between boluses. These supplements to the grassy diet included saltbush, sagebrush, water birch, and blue spruce. Although the dominance of grass and sedge paralels the diet of the woolly mammoth, tree- and shrub-browse probably played a larger role in that of the Columbian mammoth.

Two further lines of evidence are increasingly being used to help determine the diet of extinct mammals (see p.173). In the first, the microscopic wear on the surface of tooth enamel, produced during chewing, is examined under the scanning electron microscope. Animals that eat mainly grass tend to show a predominance of scratch marks on their teeth, while those accustomed to browsing the softer leaves of trees or shrubs show a greater proportion of pits. A molar of steppe mammoth from the Campo del Conte site in Italy, for example, showed three times as many scratches as pits, implying a predominantly grassy diet.

The second line of current research is the analysis of isotopes (see p.173)—variants of common elements that differ slightly in their atomic weight. Examination of the isotopic composition of mammoth protein compared to that of other Ice-Age mammals indicated a relatively high proportion of the rare $^{15}$N isotope of nitrogen, suggesting that it tended to eat older, dryer grass than horse or woolly rhino.

# FEEDING

Like living elephants, an adult woolly mammoth of 6 tons needed about 400 lb (180 kg) of fresh food a day to fuel its great bulk and may have spent up to 20 hours a day feeding. For the larger Columbian mammoth, even more food would have been required. The mammoths were superbly equipped for procuring and dealing with this quantity of food, especially with their specialized trunk and teeth.

The mammoth's main feeding apparatus, as in living elephants, was its trunk. Since all four legs were permanently occupied in support and movement, the trunk was a crucial organ, acting like a free hand for moving, breaking and manipulating all kinds of objects. It was also highly sensitive in touch, both for feeding and in social interaction.

Packed with small muscles, the trunk was capable of movement in any direction and also contraction and extension. The trunk of the Liakhov mammoth was frozen at 6 ft 6 in (2 m) in length. In baby Dima, six months old, the trunk measured about 2 ft 6 in (76 cm).

The most interesting aspect of the mammoth's trunk was its tip. In 1924 the end of a trunk was found frozen in the Middle Kolyma region of Siberia, complete with its two long, fingerlike projections at the tip. The front one is about 4 in (10 cm) long and

*A unique quantity of mammoth dung has been found at Bechan Cave in Utah (left), including this complete bolus (above), about 8 in (20 cm) across. Dissection of the dung balls gives a direct insight into the Columbian mammoth's diet.*

relatively narrow and pointed. The back one is broader and about 2 in (5 cm) long. Exactly the same structures were seen, smaller of course, when baby Dima was excavated in 1977 (see p.82).

In feeding, the trunk was used in two main ways. First, the whole trunk could be wrapped around large tufts of grass, which would be pulled up or broken off before being stuffed into the animal's mouth. Where

## "Mammoths may have spent up to 20 hours a day fueling their great bulk"

trees or shrubs were available, the trunk would have been used to break off leafy branches. Second, the "finger and thumb" at the end of the trunk could be used for delicate operations. These would have included picking flowers or buds, but perhaps also eating the relatively short grass of the mammoth steppe.

*Microscopic scratches* on the surface of mammoth tooth enamel are due to tiny particles of silica in the animal's grassy diet, as in this specimen from Campo del Conte, Italy.

*The trunk tip of* the Middle Kolyma adult shows clearly the "finger and thumb" projections used in feeding. This tip was hacked from the carcass, so the two breathing tubes (nostrils), which ran the length of the trunk, can be seen at the cut surface.

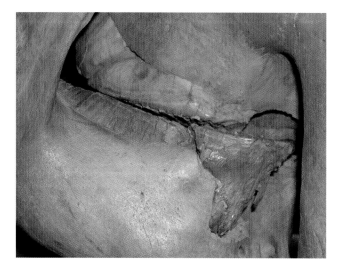

*The replacement of one tooth* by another is clearly shown in this close-up of a mammoth skull from NE Siberia. At the front of the jaw (to the right of the picture) the remnant of a worn-out molar is about to be discarded, while the next in the series is already in place in the jaw behind.

Once inside the animal's mouth, the food was attacked by the mammoth's great molar teeth. Grass is a particularly tough food, and given the quantity that the animal needed to process each day, the teeth had to be wear resistant. This requirement was all the more necessary because of the grit that the animal inevitably picked up with its food.

Each molar tooth comprised numerous ridges of hard enamel, separated from each other, and held together by dentine and cement. As the teeth ground together, the sharp enamel ridges on the upper and lower molars cut past each other to break up the food. Because the enamel ridges ran from side to side in the jaw, they cut past each other when the animal moved its jaw back and forth. This contrasts with deer or cattle, where the enamel ridges run from back to front, and the main chewing movement is sideways.

Like living elephants, a mammoth went through six sets of teeth in its life. At any one time only one tooth was fully in operation in each of the four jaws (upper and lower, left and right). As that tooth wore out, it was replaced from behind in a sort of continuous conveyor belt system, the process being repeated five times. This contrasts with humans, where each tooth is changed only once (in childhood), and the replacement grows from below.

In step with the animal's growth, successive molar teeth were progressively bigger and had more enamel ridges. The first tooth, already erupted at birth, was only ½ in (13 mm) long and lasted about 18 months. However, by six months the second tooth had already started to erupt behind. This was followed by the third tooth, which lasted until about 10 years of age. The process continued for about 30 years, when the sixth

## CHEWING POWER

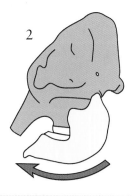

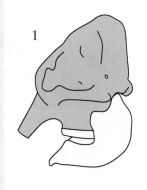

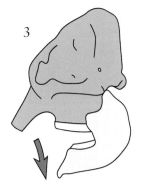

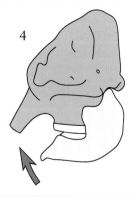

Powerful muscles operated the jaw. In the first stage, food was held between the molars. In the second, the lower jaw slid forward, chewing the food. In the third, the lower jaw opened and was pulled back, while in the fourth stage it closed, ready to move forward again in another chewing stroke. As the lower teeth ground against the upper teeth, which were fixed motionless in the skull, the hard enamel ridges cut across each other, breaking up the plant food.

and final molar came into play, the largest of the series and lasting the animal for the rest of its life. These teeth were commonly over 1 ft (30 cm) long and 4 lb (1.8 kg) in weight.

Only 1 in (2.5 cm) or less of each molar actually projected above the gum; below it the crown was up to 9 in (23 cm) deep. As well as moving forward, the tooth was also being pushed up to expose more crown as the chewing surface wore down. These high crowns further enhanced the chewing life of each tooth. When the last molar finally wore down to the root, the animal could no longer chew, and this would have caused the demise of animals that had reached an advanced age.

*These upper molars of a woolly mammoth show the great difference in size between baby and adult teeth. The first molar (above center), with which the animal was born, is no larger than a human molar, while the third of the series (above left) is about 6 in (15 cm) long. The sixth molar (above right), is a foot (30cm) long.*

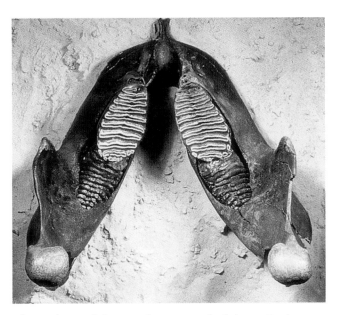

*The teeth reveal the age of a mammoth skeleton. In the woolly mammoth jaw from Condover, England (above), seen from above, the sixth molars are beginning to replace the fifth set from behind, indicating that the animal was about 30 years old at death. The jaw from Polch, Germany (bottom), shows the sixth molars almost worn out. The animal, at least 60 years old, may have died through reduced ability to feed.*

## TOOTH ENAMEL

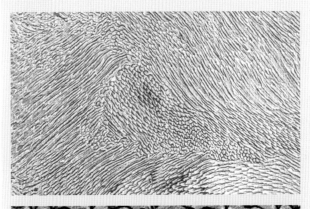

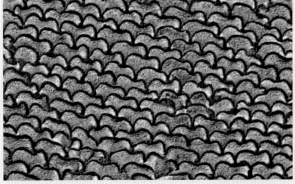

Microscopic analysis of tooth enamel is shedding further light on the mammoth's feeding adaptations. Under high magnification, mammalian tooth enamel is seen to consist of closely-packed fibres called "prisms," that can run in any direction (top). In the course of the mammoth's evolution, a greater proportion of these became directed toward the chewing surface, providing maximum resistance to wear. The lower picture shows the prisms in cross-section.

# THE MAMMOTH'S TUSKS

After its woolly coat, the mammoth's tusks were its most conspicuous feature. The world record woolly mammoth tusk, from the Kolyma River, measures 13 ft 7 in (4.2 m) along the curve, and weighs 185 lb (84 kg). Weights of very large tusks in excess of 200 lb (91 kg) have been reported. More typical is a length of 8–9 ft (2.4–2.7 m) and a weight of about 100 lb (45 kg). These figures apply to males; in females, the tusks were considerably smaller and also thinner and less tapering. Typical female tusks were about 5–6 ft (1.5–1.8 m) long and weighed only 20–25 lb (9–11 kg).

The tusks of the Columbian mammoth were as large as or larger than those of their woolly cousins. Celebrated examples include the skeleton from Jonesboro, Indiana, in the American Museum of Natural History in New York, whose tusks measure 11 ft 6 in (3.5 m), and the Franklin County skull, Nebraska, with tusks of at least 12 ft 6 in (3.8 m). A tusk from Mexico measured 13 ft 9 in (4.12 m), but the world record is in a skull of a Columbian mammoth from Post, Texas, measuring more than 16 ft (4.9 m). It was presented to the American Museum of Natural History in 1934.

The massive tusks were preceded in baby mammoths by tiny milk tusks. Only a couple of inches long, these erupted at about six months and lasted about a year. They were replaced by the permanent tusks which grew continuously through

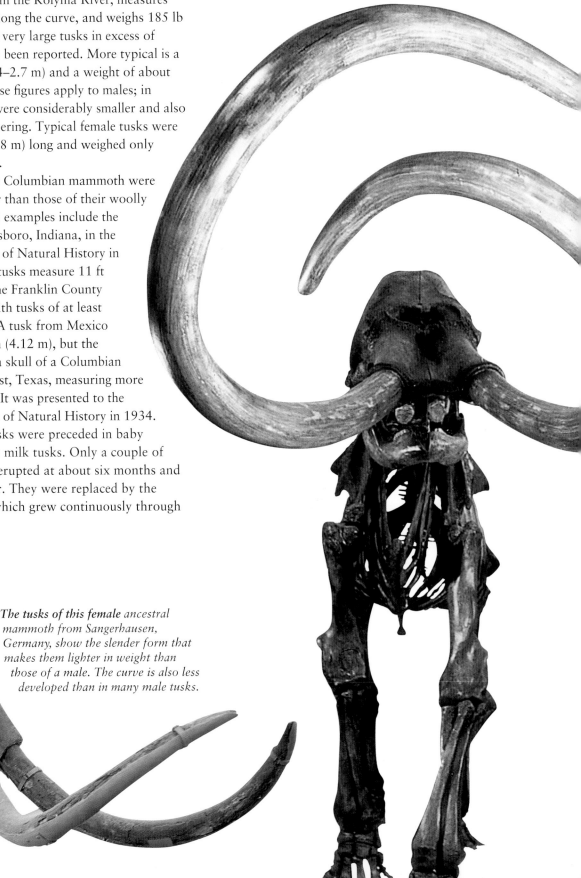

*In front view a mammoth's tusks* were an imposing sight, as dramatically shown in this male skeleton of a Columbian mammoth from the tarpits of Rancho La Brea, California (see p.68).

*The tusks of this female* ancestral mammoth from Sangerhausen, Germany, show the slender form that makes them lighter in weight than those of a male. The curve is also less developed than in many male tusks.

# TUSK WEAR AND TEAR

Despite the hard, durable nature of ivory, tusks often show signs of wear and stress. The lower sides of the end of the tusks usually became flattened, polished and scratched (below left) through abrasion against the ground during feeding. The end could also be snapped off (below right), perhaps during fighting. In this example, the shattered tip has been polished smooth by subsequent wear.

life, although the rate of growth slowed down in adult animals. Up to a quarter of the tusk length was embedded in the socket in the skull, and it was here that new tusk material was added, at a rate of between 1 and 6 in (2.5 and 15 cm) per year.

The shape of mammoth tusks was different from that of living elephants. Mammoth tusks have been described as having a "twist" or a "writhe." They grew in the form of a spiral or corkscrew, in some cases quite tightly, in others very openly. The left and right tusks twisted in opposite directions and occasionally formed almost complete circles, with the tips crossing in the middle. It has been suggested that this three-dimensional curvature allowed the tusks to grow very long and powerful, but with less torque on the skull than if they had been straighter, since the mass was held closer to the head. However, in general, the tusks of the Columbian mammoth seem less twisted than those of its woolly cousin.

As in the case of living elephants, the mammoth's tusks were undoubtedly used for a variety of activities. A key use was in intimidating, sparring and fighting with other individuals. Males would have fought for access to females, and both sexes would have had occasional tussles over favored feeding spots. Such contests would primarily have involved pushing and twisting as a trial of strength. The markedly curved tusks would have been unsuitable for stabbing the rival, the aim of serious fights among elephants today. An alternative may have been to crash the tusks down on the rival's back, as seen also among living elephants, and as suggested by broken mammoth shoulder blades (see p.110). The large curved tusks of adult males may have acted as a sexual attractant to females and would have been a deterrent to predators.

## "The largest mammoth tusk in the world is more than 16 feet long"

Tusks were also used in feeding. Mammoth tusks very frequently show areas of wear, sometimes forming a flat, polished facet up to 1 ft (30 cm) long, usually on the side of the tusks that would have touched the ground. This type of wear has led to the theory that the tusks were used as a snow plough, the animal moving its head from side to side to clear away snow and expose vegetation beneath. Although this may have been an occasional use, it must be remembered that snowfall was not a serious problem over much of the mammoth's range. Moreover, modern elephants also often show wear and scratching near the tips of their tusks. These result from stripping bark and digging up plants, activities that mammoths would have undertaken, especially when trees or shrubs were available. It has also been suggested that an important function of the tusks was to break up ice to eat in winter, when there was no accessible unfrozen water to drink. If so, this practice would certainly have contributed to wear on the tusks.

*The tiny milk tusk of a woolly mammoth, shown at life size. To the left is the root; to the right the plug enamel at the tip. This specimen is from Kent's Cavern, Devon, England.*

THE NATURAL HISTORY OF MAMMOTHS          95

# FIGHTING AND MATING

Two male mammoths fight for access to a female in this Siberian setting during the last Ice Age, about 18,000 years ago. Other mammoths gather excitedly around, while in the background a third male takes advantage of the situation to begin courting the disputed female.

*With a gestation period of 22 months, woolly mammoths may have mated mainly in the summer to ensure birth in the spring nearly two years later. Then, lactating mothers would be supported by fresh spring growth. Only females in breeding condition (oestrus) would be attractive to the males, which themselves may have undergone cycles of aggressive behavior related to mating.*

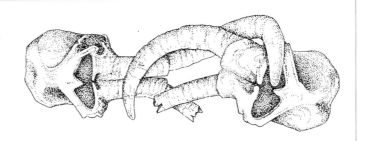

*A **unique fossil find** was unearthed in Nebraska:* two mammoths that had died with their tusks interlocked. Mammoths could turn their curved tusks into deadly weapons, by thrusting them upward, swiping them sideways, or crashing them down from above. Interlocked, they provided leverage for a twisting and pushing contest, which could occasionally have resulted in the breakage of a tusk.

***Mock or ritualistic sparring**, as in living elephants, was probably common among male mammoths and would have involved head butting, pushing, and clashes of tusks. Just displaying his tusks would often have been enough for one male to assert his superiority. Occasionally, however, serious fights could develop, especially between evenly matched individuals pursuing a receptive female. These were violent and could sometimes result in the death of one or both combatants.*

# MOTHERS AND YOUNG

A family group of woolly mammoths would usually have been composed of up
to a dozen adult females and their young, which formed a closely knit social unit.
This Ice Age scene is set in northern Siberia during the spring, when a young
mammoth has fallen through thin ice and is helped to safety by its mother.

A female mammoth suckles her baby while caressing its back with her trunk (below, right). Unusually among mammals, mammoths (like living elephants) had their mammary glands toward the front of the body. This was the only time in a mammoth's life when it took food directly with its mouth (right). Some plant food was taken from a few months of age, and the animal was weaned by the age of two or three, by which time the mother would be ready to breed again.

The matriarch, an old and experienced female and leader of the group, stands to the left, surveying the scene. The other adults are likely to be her sisters, offspring, or nieces. If any individual became sick or injured, others would offer constant assistance. If a mother died, her offspring would be adopted by one of the other females.

Like other juvenile mammals, mammoth youngsters were constantly at play. One animal clambers over its mother's back, while another tugs at the tail of an older cousin. On reaching maturity at about 12 or 15 years of age, young males gradually spent less time with the group before finally setting off on their own, like the adolescent seen in the background of this scene.

## LIFE CYCLE AND BEHAVIOR

All elephants are highly social animals, and mammoths were no exception. It may seem surprising that social behavior can be deduced for animals known only from bones and carcasses. However, certain fossil assemblages, together with Ice Age art and our knowledge of modern elephants and the mammoth's environment, allow the broad features of its life cycle and behavior to be outlined with some confidence. Studies of growth rings in tusks are also starting to provide valuable clues (see pp.106–107).

> ## "Evidence suggests that mammoths shared patterns of behavior with living elephants"

Elephant society is matriarchal: it is organized into family groups led by adult females, with one experienced female, the matriarch, dominant. A family group may comprise anything from 2 to 20 or so individuals, with the average group size being about 10. Adults within a group are relatives: mothers and daughters, sisters, aunts, and nieces. They are accompanied by their offspring of

*A drawing in the cave at Rouffignac, France, seems to suggest that mammoths shared the close social bonding so evident in elephant families. It shows a meeting between two small groups, similar to the renewal of bonding that occurs when related elephants meet after a period of separation.*

both sexes, up to the age of about 10. Above this age, young females stay within the family group, but males start to move away. Adult males live singly or in small bachelor groups, associating with females mainly for purposes of mating.

Apart from the likelihood that, being close relatives, mammoths shared a similar social structure to living elephants, there is some indirect evidence for it. At the fossil site at Dent in Colorado, for example, the remains of Columbian mammoths seem to comprise entirely females and young. It is debated whether they met their death naturally or at the hands of prehistoric people (see p.149), but in either case a female-led family group is implied. Other examples of assemblages dominated by adult females and young, apparently a natural group that met a sudden death, include those from Sevsk, Russia, and Waco, Texas (see pp.62–3), and another site in Russia, Teguldet

**The close kinship of mammoths** to living elephants (left) makes it likely that they shared many aspects of behavior. Here two male African elephants engage in a ceremony of greeting, recognition, and assessment. Their ears are limp, indicating no serious aggression, although such activities may sometimes develop into shoving and butting matches.

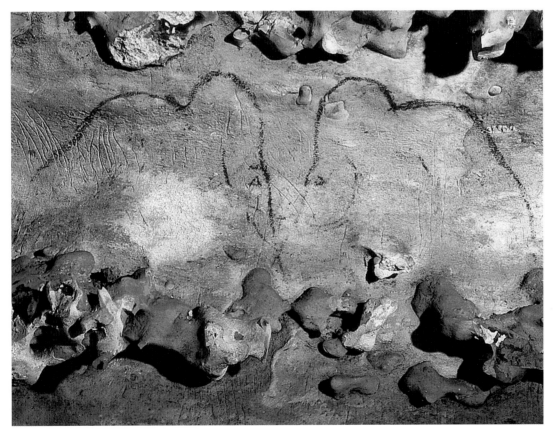

**Two mammoths meet face to face.** This drawing in Rouffignac Cave could represent either a greeting or confrontation.

near Tomsk. Here, 24 mammoths, including six skeletons, were found during sand quarrying, and have been dated to around 22,500 years ago.

At other sites, the opposite is found. For example, at Hot Springs in South Dakota all except one of the numerous skeletons is male. This site was a sinkhole forming a natural trap into which animals fell (see pp.64–5). With modern elephants, it is usually single, roving males that are the more adventurous and more likely to get into difficult situations, while the females, living in groups, are more cautious and help each other out of trouble. The Hot Springs site has been interpreted as an accumulation of individual male deaths over a long period, implying that, as in living elephants, male mammoths roamed alone. Moreover, most of the animals were in their early adult life—the time when male mammoths were probably at their most active and adventurous.

The head of the Yukagir mammoth, the most complete ever found, provides remarkable new evidence on mammoths' behavior that links them to the living elephants. On the side of the face, about half-way between the ear and eye, is a small opening. This is the outlet of the temporal gland, an organ that in living elephants, males especially, produces a strong-smelling, oily liquid called temporin. Roughly once a year, bull elephants enter a phase of several weeks or months called musth. During this time they are highly aggressive and sexually active, while paying less attention to feeding. A clear sign of musth is the constant dribbling of temporin down the side of the face. The discovery of the temporal opening in the woolly mammoth strongly suggests that males in this species, too, underwent periodic musth, the hairy covering on the side of its face only serving to aid the aerial dispersal of the scent, signaling the bull's condition to other members of the herd.

The mammoth trackways uncovered at St Mary's Reservoir, western Canada, in 1999 (see pp.73–4), provide a different kind of remarkable insight into mammoth behavior. The trackways, ranging in diameter from 5½ in to a huge 23½ in (14 cm to 60 cm), cover an approximately equal mixture of juveniles, subadults, and adults. Adults walked with stride lengths of up to 6½ ft (2 m). Meanwhile, a calf ran to keep up with its mother, and two juveniles curved around, chasing each other.

*A "mammoth family"* reconstructed from the skeletons found at Teguldet, near Tomsk, Russia. An adult female (right) accompanies an adolescent female (left), and infant.

*The head* of the Yukagir mammoth, the most complete ever recovered, shows the complete ear (right), eye, and mouth (below).

It is very difficult to estimate the number of individuals of a species that lived at any one time in the past. Abundance in fossil deposits does not necessarily equate to large numbers of living animals, as the assemblage may have accumulated over a long period of time. Estimates of mammoth density have varied greatly among researchers, from 1 to 150 individuals per 10 square miles (25 square km). In reality we have little idea, but the figure probably varied considerably with the seasons and with the animals' life-cycle events.

Among elephants, larger herds are sometimes formed as family groups come together. Before the great reduction in elephant numbers in recent decades, gatherings of up to a thousand elephants were occasionally seen, usually for migration to new feeding grounds. It is probable that mammoths also sometimes formed large herds; indeed, there are reasons for thinking that this was more prevalent than in the living elephants. It is generally found that large animals living in open country are more likely to form herds than related species in woodland or forest. Examples are seen in the herds of antelope on the African plains, in contrast to the comparatively solitary existence of their forest-dwelling relatives. Mammoths, especially woolly mammoths, lived in a largely open, treeless habitat, so herding may often have occurred.

*Trackways of the woolly mammoth, uncovered by accident when a reservoir in western Canada was drained in 1999, provide a remarkable example of "fossilized behavior."*

*The tiny opening of the temporal gland, seen for the first time in a mammoth, provides vital clues to the animal's behavior.*

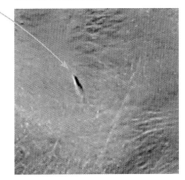

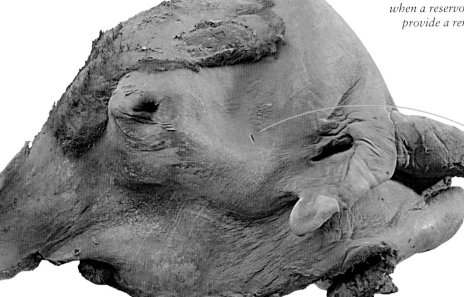

The formation of large herds would also have been likely if mammoths made long seasonal migrations in search of food. The northern mammoth steppe provided a rich source of vegetation during the spring and summer when plant growth was at its peak. In the fall and winter months, however, fresh food would have been scarce, and herds may have migrated south. As yet, however, there is no clear evidence for long-distance migrations among mammoths. An ingenious study that measured strontium isotopes in the bones of Columbian mammoths from Blackwater Draw, New Mexico, deduced that the animals must have spent part of the year in the Rocky Mountains, implying migrations of around 124 miles (200 km). New research on tusk rings (see pp.106–7) promises to provide an insight into this question.

The northern habitat of the woolly mammoth may have had a direct effect on its life cycle. In common with other tropical animals, modern elephants breed at all times of the year. However, on the African savanna many individuals time their breeding so that the young are born in the rainy season when food is most plentiful. Mammoths lived in northern climes during the Ice Age, when the variation between summer and winter was more intense than in the modern tropics. It is therefore quite likely that their breeding was strictly timed so that young were born only during the plant growth season.

A similar pattern can be seen, for example, in today's deer from Europe or North America, which give birth in the spring, in contrast to tropical deer from South America and Southeast Asia, which breed all year round. A recent study of various baby mammoth carcasses has confirmed this idea, suggesting that they were all born in the spring and summer. Like living elephants, the gestation time of mammoths was probably 21–22 months, so a mating season in the summer to fall, producing young in the spring and summer nearly two years later, can be calculated.

The total lifespan of different mammoth species can be estimated from their size. Among mammals, there is a close relationship between a species' size

*A "lunar landscape" was exposed at the site of Murray Springs, Arizona, excavated in the 1960s. Several skeletons of Columbian mammoths were discovered, together with stone implements. In one area, removal of sediment revealed a series of depressions identical to modern elephant footprints in soft ground. The tracks appear to lead up to one of the skeletons.*

*At Waco, Texas, a group of female and young mammoths were arranged in a circle, leading to speculation that they had formed a defensive ring when overwhelmed by floodwater.*

## DANGER FROM PREDATION

Spotted hyenas, seen here scavenging the carcass of an African elephant, lived in Europe during the Pleistocene. At Kent's Cavern in southwestern England, a former hyena den, thousands of hyena fossils have been excavated, mingled with bones of juvenile woolly mammoths, many of them chewed.

At Friesenhahn Cave in Texas, juvenile Columbian mammoth bones were accumulated by the now extinct scimitar-tooth cat, *Homotherium*. Neither hyenas nor big cats could have hunted mammoths in their prime, but they may have picked off young animals.

*Above, a hyena scavenges the carcass of an African elephant. Below, at West Runton, England (see p.75), hyena droppings were found, some of them (arrowed) in upright position where they fell, among the remains of a 700,000-year-old steppe mammoth. Several bones bear marks of hyena chewing.*

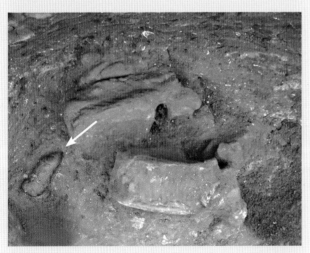

and its longevity. On this basis, woolly mammoths, at around 6 tons, probably lived as long as living elephants—about 60 years. Columbian mammoths, which frequently reached 10 tons, presumably lived longer—perhaps up to 80 years or more. Study of tusk growth rings may resolve this question (see p.106–7).

At Murray Springs in Arizona an extraordinary piece of "fossilized behavior" was discovered. A trail of buried footprints was uncovered which can only have been produced by an elephant, and the only possible local candidate is the Columbian mammoth. Moreover, the footprints were found to lead up to a fossilized skeleton of that species. Two explanations are possible. Either the footprints belonged to the preserved individual and represent its last moments alive; or they may belong to

> ## "Modern elephants spend days guarding the corpse of a dead relative, even 'burying' it with leaves or soil"

another individual which came to help the dying animal, or mourn its death. Modern elephants spend days guarding the corpse of a dead relative, even "burying" it with leaves or soil. It is not certain that mammoths showed the same traits, but their close relationship to living elephants makes it a distinct possibility.

A further intriguing clue to mammoth behavior is provided by rocks that seem to have been rubbed by the animals. In 2001, on California's Sonoma Coast, 50 miles north of San Francisco, some blueschist and chert outcrops were observed to have highly polished areas, some of them 10 to 13 ft (3 to 4 km) above ground. They occur on vertical faces but particularly at rock edges. In Africa rubbing rocks are fairly common in the savanna and grassland areas. African herbivores—elephant, rhino, zebra, and buffalo—rub against rocks, tree trunks, and termite mounds, usually after bathing, to rub off mud and ectoparasites. If the Californian rubbings were indeed produced by large herbivores, mammoths or mastodons are the only species large enough to have rubbed at that height, and the site has been named "Mammoth Rocks." Similar rubbing sites have been known since the 1940s at Hueco Tanks, Texas, and Cornudas Mountain, New Mexico.

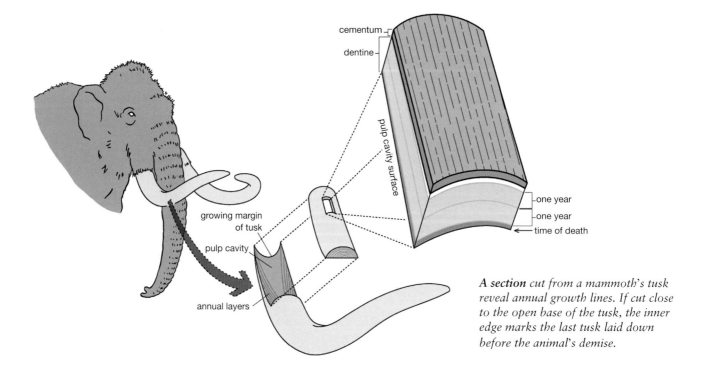

cementum
dentine
pulp cavity surface
one year
one year
time of death

growing margin
of tusk

pulp cavity

annual layers

*A **section** cut from a mammoth's tusk reveal annual growth lines. If cut close to the open base of the tusk, the inner edge marks the last tusk laid down before the animal's demise.*

## CLUES FROM TUSK RINGS

A mammoth's tusks grew throughout its life. At the base of the tusk, deep inside its socket in the skull, new tusk material (dentine) was laid down, adding to the length of the tusk which was at the same rate pushed out from the socket. The mineral was laid down on the inner surface of the cone-shaped pulp cavity that usually extends to around the level of the socket rim, so that the tusk beyond this is solid. A section cut lengthways through the tusk shows the ivory as a long series of cone-shaped sections, reflecting its deposition around the edge of the pulp cavity. In cross-section, these appear as a set of concentric rings.

The total length of an adult mammoth's tusk represents many years of growth, in some cases virtually its entire life-span. Moreover, growth lines can be seen in the tusk, forming a kind of "medical record card" of the animal's life. The major growth lines are annual, and between them can be seen finer lines, usually weekly. Under higher magnification still, daily lines are

visible, each comprising approximately 0.015–0.025 mm of dentine deposition. Research by Dan Fisher and his colleagues has begun to unlock the secrets hidden within the tusks. A plug or cube is cut through the full width of the tusk, and the cut surface, or a thin section, is viewed under the microscope. Pigmented resin can be used to show up the rings. By sampling at intervals along the tusk, a complete record of the tusk's growth is obtained.

Counting the total number of annual rings in the tusk gives the length of time in years that the tusk grew. In principle this might allow us to determine the typical longevity of a mammoth, but in practice this is difficult because the tusks wore down in life through use, so the early years of life (at the tip) will be missing. The maximum number of annual rings so far recorded in a single tusk is 42. Each annual ring generally comprises a light and dark band, the dark band corresponding to the winter months. This has been confirmed by

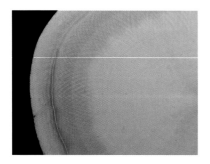

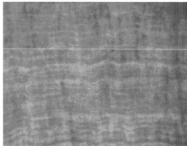

*A **cross-section** through a mammoth's tusk (far left) shows annual rings of ivory (dentine) growth—about six are visible in this specimen. Under high magnification, thin bands of daily growth are visible (left) within each annual layer.*

A **mammoth's tusk** is sampled (left) with a hollow drill that removes a plug (center) covering several annual growth rings. A view of the open end of the tusk (right) shows the pulp cavity that in life was buried in the skull. Dentine (ivory) was laid down on its inner surface, and as the cavity filled up it lengthened by addition to its edges, pushing the tusk out from the skull.

measuring changes in oxygen isotope proportions along the tusk, varying roughly with the seasons on an annual cycle (see p.172). As a result, it is possible to determine the season at which the mammoth died, by examining the last-formed dentine, at the surface of the pulp cavity.

Analysis of a number of North American mammoths has shown that they often died in the late winter or early spring, usually the most difficult time for northern animals. Sometimes there is a reduced growth rate towards death, perhaps due to starvation or disease. In the Jarkov mammoth (see p.57), analysis of nitrogen isotopes suggested brief periods of nutritional stress each year in the late winter, although this reversed with the onset of spring. It was also possible to show, from the oxygen-isotope profile, that the animal had ingested winter precipitation, presumably in the form of snow.

There is evidence, too, that the major climatic cycles of the Ice Age impacted the animals. Slower growth is seen in mammoth tusks dated to the interval around 25–20,000 years ago, the most severe phase of the last glaciation. The tusk studies may also shed light on the long-standing question of whether mammoths undertook seasonal migrations. Analysis of isotopes across several years of the Jarkov mammoth's life does suggest periodic changes in the animal's feeding environment, but these appear to be of irregular duration rather than strictly seasonal, the animal moving between more southern regions with more

moisture and vegetation cover, and more northern, open, drier habitats, where it finally died.

Researchers are attempting to pinpoint other key life-cycle events in the record of the tusks. The point at which a young mammoth was weaned can be observed in the isotope proportions of nitrogen and carbon, since these differ according to whether an individual was eating plant- or animal-derived food. Suckling its mother's milk was the only time of its life when a mammoth was eating principally animal-derived food. The team found an interval of a few years in a young mammoth's tusk rings where the isotope ratios gradually changed from a milk-fed to a vegetation-fed signature. This indicates that weaning was a protracted process mammoths, as it is in elephants.

Another key moment in a mammoth's life, at least for males, was when they reached sexual maturity and left the matriarchal group to begin an independent life. The fossils show a reduction in growth rate of the tusks at a time that can be interpreted as this interval of the animal's life, suggesting that they suffered a period of hardship away from the maternal group until they gained experience at independent foraging.

Evidence of changes in the mammoth's life-cycle, deduced from tusk growth, have been used in an attempt to determine whether mammoths became extinct due to climate change, or at the hand of human hunters (see Ch. 5).

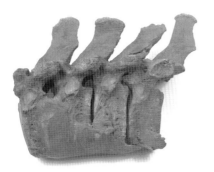

*Mammoths commonly suffered from back problems.*
*These vertebrae (above), from the Huntingdon (Utah)*
*mammoth, are abnormally fused together.*

*No two mammoths had exactly the same tusk shape, but in*
*this 16,700-year-old woolly mammoth skull, from Bzianka*
*in Poland, the left tusk (on the right) spirals down in a most*
*unusual way. The right tusk had snapped during life, and its*
*lower end had become flattened and polished through wear.*

## HEALTH AND DISEASE

Mammoths were not immune to health problems, as can be seen from various traces of disease preserved in bones, teeth, and tusks. A survey of nearly 2,000 mammoth fossils showed that about 4 percent had some signs of illness. These diseases can be identified by comparison with veterinary studies of present-day species.

One of the most common abnormalities seen in mammoth fossils is distortion of the molar teeth, due mostly to disruption of the process of tooth replacement. If the normal forward progression of the teeth was interrupted, the teeth coming from behind could get squashed into unusual shapes during their growth. Several such teeth have been recovered as fossils; the fact that many had subsequently come into use indicates that normal tooth replacement had begun again and the animal had survived, at least for a while.

### "Fossils show that some mammoths suffered from severe tooth decay"

A few teeth show evidence of periodontal disease or dental decay. Periodontal disease results from build-up of bacteria between tooth and gum, dissolving away part of the surface mineral of the tooth. A cavity (caries) may develop, which in advanced cases leads to further infection and an abscess at the base of the tooth. Among a sample of mammoth teeth from Ice Age Britain, about 2 percent showed periodontal disease, of which about half had proceeded to a caries. The relatively low sugar content of the mammoth's diet (grass rather than fruits or berries) may have helped to reduce the prevalence of these diseases.

Occasionally cancerous growths are seen in teeth, resulting in abnormal outgrowths of dental tissue. These are more often found in the fossil state than bone cancers because the latter so weaken the bone that preservation is unlikely.

A remarkable abnormality occasionally seen in mammoth jaws is the growth of an extra set of teeth. Mammoths typically went through six sets of teeth in their life (see p.92), but in some cases, a seventh formed in the jaw behind the sixth set. This so-called supernumerary molar looked in some cases like a shortened version of the sixth tooth, and sometimes came into wear, thereby potentially prolonging the animal's life beyond its usual span. In other cases, the seventh tooth was just a very small, compacted bundle of a few enamel plates.

Being organs that continue to grow through life, the tusks were susceptible to growth defects of various kinds. The two tusks of an individual were rarely perfectly symmetrical. It is a matter of judgment at what point an unusual morphology is regarded as abnormal.

The most common bone disease, found in about 2 percent of mammoth bones, is osteoarthritis. This results from the wearing away of the cartilage between joints, followed by abnormal growth of the bone ends and their possible fusion. A striking example is seen in the woolly mammoth skeleton from Praz Rodet, Switzerland, where several vertebrae in the lower back are completely fused together. The Yukagir

*In this mandible from Otterstadt, Germany, an extra set of teeth has formed behind the sixth (normally last) pair.*

mammoth suffered from a condition called ankylosing spondylitis in the 4th and 5th thoracic vertebra. The inflammation would have caused pain, especially in the early stages of abnormal bone growth. Perhaps the most striking example is seen in the Huntingdon, Utah, mammoth. In this very old animal, nearly every vertebra is severely deformed with arthritic disease, and as at Praz Rodet, there is pathological fusion of four lumbar vertebrae.

A few bones show evidence of bacterial infection such as osteomyelitis, in which bacteria gain entry to the bone tissue, resulting in abnormal outgrowths of bone. This condition could have become fatal if the infection had spread throughout the body.

There are several examples of mammoths having broken a bone, but surviving with the fracture healed.

*Because of a "log jam" in the forward progression of teeth, this molar (right), belonging to an ancestral mammoth, was pushed into a curve. The molar of a woolly mammoth (above) has a large root cancer and abscess in the left-hand side.*

One bone which may have been particularly prone to fracture was the shoulder blade, a relatively thin, flat structure. In one specimen from Siberia, the whole back part of the right shoulder blade had been snapped off and had healed in roughly the natural position.

A similar fracture is seen in the left shoulder blade of the mammoth from Condover, England, which had been broken along the top edge and knitted together with extensive growth of new bone. As a result, the healed blade in the region of the fracture is many times thicker than normal. There are numerous possible explanations for such fractures, among which are falling down a hole or being tusked by another mammoth. A possible example of the former scenario is provided by the Beresovka carcass (see pp.52–53), which may have broken its right shoulder blade on falling to its death.

Perhaps the most striking pathology was observed in the right hind-limb of the steppe mammoth from West Runton, England (see p.75). The lower end of the thigh bone (femur) was grossly deformed. Normally smooth surfaces for articulation with the shin bone (tibia) were either absent or severely distorted. Attachment points for ligaments were also lost. The articular surfaces had, however, become re-formed, albeit in a grossly

## A FRACTURED SHOULDER

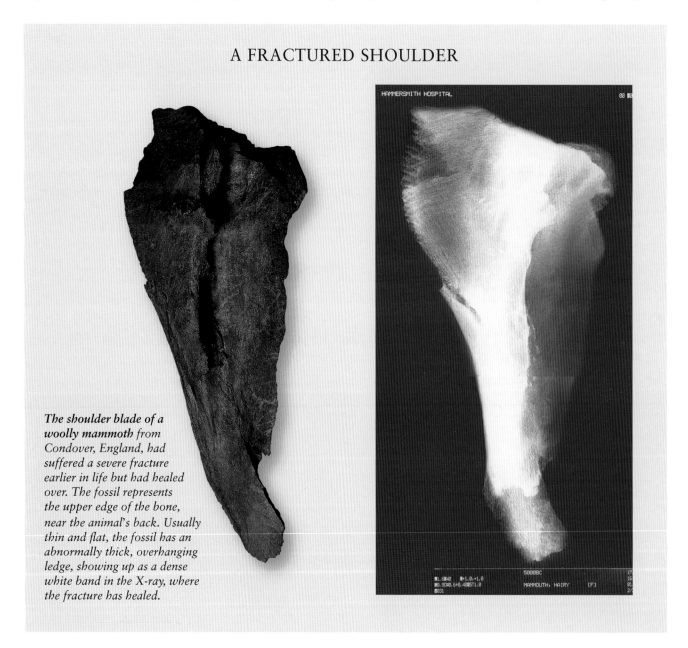

*The shoulder blade of a woolly mammoth* from Condover, England, had suffered a severe fracture earlier in life but had healed over. The fossil represents the upper edge of the bone, near the animal's back. Usually thin and flat, the fossil has an abnormally thick, overhanging ledge, showing up as a dense white band in the X-ray, where the fracture has healed.

THE NATURAL HISTORY OF MAMMOTHS

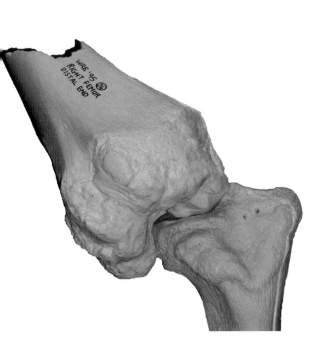

*The left and right knee joints of the West Runton mammoth. The right femur (seen on the left) had been severely dislocated.*

abnormal way, in such a way as to suggest that the joint had been used for a considerable period after the injury. The left limb bones were more normal, but the left pelvic bone and thigh bone show thickening and wear indicative of increased load, as the animal placed less weight on the injured right leg.

The animal's right tibia has not been found, so the precise form of the injured joint is uncertain, but the form of the femur strongly suggests that it had slipped severely into a lateral position, a condition technically known as subluxation. The cause of such an injury can only be conjectured, but if the mammoth fell heavily, either accidentally or during a fight, its 10-ton mass could have been sufficient to rupture and disarticulate the joint. The result would undoubtedly have been severe pain, swelling and internal bleeding, but the mammoth evidently survived the injury and retained a degree of mobility. However, its lameness would have compromised its ability to climb a muddy bank or regain its feet after a fall. This could well have contributed to its demise in the shallow waters of the river at West Runton.

## THE SOLLAS "EGG"

Abnormal growths on tusks occasionally resulted from damage to the tissues that lay down tusk mineral in the socket. A famous example of this came from excavations at Paviland Cave in Wales (see p.116). In about 1820 Dean William Buckland discovered, among a quantity of mammoth ivory, a fragment of tusk with an unusual, hollow cavity. He perceptively guessed that this was an abnormality—similar to that seen among modern elephants where the growth organ inside the tusk socket has been damaged.

Nearly a century later, during renewed excavations in the cave, W. Sollas unearthed an egg-shaped lump of ivory which fitted into the hollow of Buckland's specimen. This ivory "egg" had formed as part of the abnormal growth of the mammoth's tusk. Moreover, it had been discovered by prehistoric people, who had pierced a hole in it to make it into a pendant.

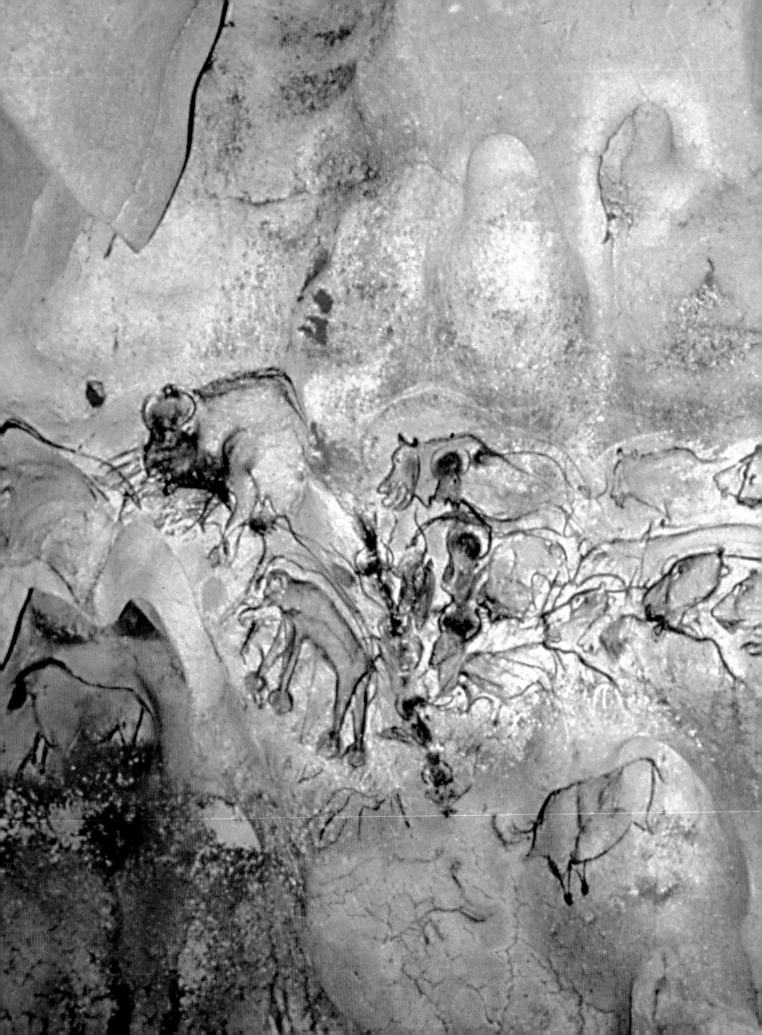

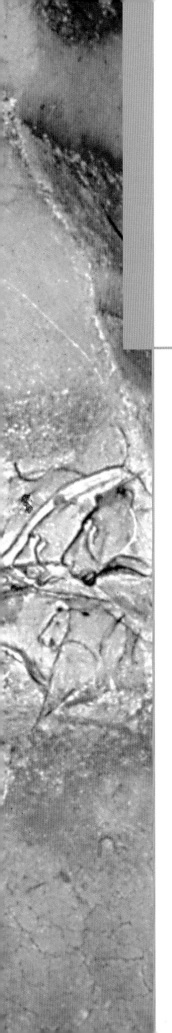

# MAMMOTHS AND HUMAN CULTURE

Of the many extinct animals that once roamed the Earth, mammoths are among those which lived at the same time as humans, often even alongside them. The vivid, lifelike depictions of mammoths in cave paintings, and the ever-growing archive of evidence of the way humans made use of mammoth bones and ivory, not only prove that mammoths and humans coexisted but also help underline the importance of mammoths in the daily lives of our ancestors.

From a plentiful supply of mammoth bones and tusks—probably scavenged more than acquired by hunting—humans made anvils, tools, and even hut-style dwellings. Prehistoric sculptors also recognized the beauty of ivory and fashioned it into highly original and sophisticated carvings, which are among the first of their kind created by human hand. And today, thousands of years later, fossil mammoth ivory is still being crafted to make jewelry and ornaments.

*The "Lion Frieze" in France's Chauvet Cave, discovered in late 1994; two superimposed mammoth figures can be seen, the larger being in outline, and the smaller one more detailed, with the eye and small ear depicted, as well as its four feet depicted as "tracks." Some of Chauvet's art has been ascribed to about 30–35,000 years ago, but this is still highly controversial and uncertain.*

# MAMMOTHS IN THE ARTIST'S EYE

Artists work in the glow of naked flames to decorate a cave in southern France, about 18,000 years ago. The large number of paintings that have survived indicate that cave decoration was a widespread activity in certain regions. Mammoths feature regularly among the animals depicted.

*Cave painters used a variety of naturally occurring pigments to create their colors. Iron oxide (hematite or ochre) produced red; manganese dioxide and charcoal produced black. Various colors, from yellow to purple, can be produced by heating ochre. It is not known which binding medium was used: experiments suggest that calcium-rich cave water was best for fixing pigments on the damp rock faces, but recent analyses of paint from several caves suggest the use of animal and plant oils.*

*Many of the caves were completely dark inside. To illuminate their work, the cave painters may have used stone lamps filled with animal oil. An engraved sandstone lamp (right) was found at Lascaux, France's most famous decorated cave. Wood fires and brush torches would also have provided plenty of light.*

**Most cave walls were within easy reach** of the artists, but in a few cases high panels or ceilings were decorated with the aid of ladders or scaffolding. One gallery at Lascaux has about 20 sockets cut into the rock on both sides, 6 ft (1.8 m) above the floor, which probably held platform joists.

**Paint was applied to the rock with the fingers or**, more often, with some kind of tool. None has survived, but experiments suggest that brushes of animal hair or of crushed and chewed plant fibers would have produced the best results. Pads covered with dampened powder were used in some caves. On rough surfaces, or to make hand stencils, artists would spray liquid paint through a tube or directly from the mouth.

## THE CO-EXISTENCE OF PEOPLE AND PREHISTORIC PACHYDERMS

In the early 19th century many paleontologists and geologists were beginning to ask questions about the antiquity of the human race, and whether it was conceivable that in the remote past people had lived alongside animals that were now extinct. Some finds had already suggested such a coexistence, but their significance had not been recognized. For example, in about 1690 a piece of pointed black flint was found close to some elephant bones in a gravel pit in Grays Inn Lane, in London, by John Conyers, a pharmacist and antiquary. It was, in fact, an early Stone Age handaxe, but at the time it was assumed to be a weapon used by a Briton to kill an elephant brought over by the Romans in the first century A.D.

> "In the early 19th century it was generally believed that humans could not have lived at the same time as extinct animals"

In 1823 the Reverend William Buckland, an Anglican priest and the first professor of mineralogy and geology at Oxford University (or indeed anywhere in Britain), published details of his discovery of an Old Stone Age male skeleton, stained with red ochre, in the Goat's Hole Cave at Paviland in Wales. Also in the cave were rhino and bear bones, together with an ivory pendant (see p.111), fragments of slender ivory bracelets, rings and rods, and lengths of tusk, and the skull from an "elephant." It has since been established that the ivory and skull were mammoth, but Buckland did not believe that humans and fossil animals could have lived at the same time, and so suggested that the "Red Lady of Paviland," as the human skeleton was mistakenly labeled, was Romano-British.

Other people digging in western Europe, however, were being led to different conclusions, since they repeatedly found tools of stone and bone sealed beneath stalagmite floors in caves or rock shelters. These tools had been made by humans and were clearly associated with the remains of extinct animals. In some finds from southern France the animal bones even bore cut-marks that could only have been made by humans.

The mounting evidence was becoming overwhelming, and international recognition of the intermeshing of human history with that of extinct animals was finally won by the work of amateur archeologist Jacques Boucher de Perthes in the Somme gravels of northern France. By 1859, the year which saw the publication of Darwin's *On the Origin of Species*, the great antiquity of the human race was accepted by all but a few diehards.

In May 1864 the French paleontologist Edouard Lartet made a find remarkable enough to convince the remaining sceptics in the scientific community. Digging in the rock shelter of La Madeleine in the Dordogne, he came upon a lifelike engraving of a mammoth on a large piece of mammoth ivory. This was not the first piece of Ice Age art to be found, but it was among the first whose true age could be proved from its careful excavation and its context, and above all it clearly depicted an extinct animal drawn on a fragment of that same species. There could be no surer proof that people had lived alongside the vanished fauna of the Ice Age.

## PORTABLE MAMMOTHS

Humans were producing art in the form of cave paintings and sculpture throughout the latter part of the last Ice Age, from about 35,000 to 11,500 years ago. Although this art is dominated by apparently abstract motifs and by images of horses and bison, the mammoth was the next most frequently depicted animal. Over 500 representations of it are known, spanning a wide range of media, from paintings and engravings in 46 decorated caves in Spain, France, and Russia, to numerous engravings on bone, stone, antler,

*A mammoth carved from antler 16,500 years ago* formed the decorative end for a spear-thrower found at Bruniquel in central southern France. The broken tail would have provided the hook onto which a spear could be mounted to provide extra leverage for throwing. The tusks are carved in low relief along the antler shaft, while the head itself is lowered and the trunk descends to the feet.

116 ▊ MAMMOTHS AND HUMAN CULTURE

and ivory, as well as three-dimensional carvings in antler and ivory ("portable art").

The discoveries of Ice Age portable art in the 1860s led to a headlong treasure hunt in caves and rock shelters in Europe. Depictions of extinct animals, or of animals that are no longer found in southern Europe, such as reindeer, were particularly prized. As a result of this wave of enthusiasm, scores of portable engravings and carvings of mammoths were unearthed, and many more have since been discovered.

Unlike depictions on cave walls or rocks, most of these portable images can be confidently dated to the Ice Age because they have been found in the same archeological layers as tools and other remnants of occupation from that period. By far the biggest collection was found in the 1960s during the excavation of an open-air camp dating to about 15,000 to 14,500 years ago at Gönnersdorf in northwestern Germany. It contained hundreds of small stone plaquettes, many of them engraved with animals or stylized women. There are at least 62 mammoths, drawn on a total of 47 plaques, ranging from less than 2 in (5 cm) to animals measuring 12 in (30 cm) from eye to tail. About half of the figures were found intact.

In common with all Ice Age examples of mammoths, the Gönnersdorf mammoths are depicted in profile and

*A mammoth engraved on mammoth ivory, found at La Madeleine in France, proved that humans once coexisted with mammoths.*

are rather static. Some are simple outlines, but others show a considerable amount of detail and provide clear evidence of how mammoths must have looked. It appears that both adults and young are shown, identified by the line of the back: in the adults this descends steeply to the tail, which is level with the mouth. The young have a domed back that is highest in the middle, while the lower part of the body is so shaggy that the stomach cannot be seen.

In the adults and young alike, the head generally has a small protuberance in front of the eyes, the ear is small and mostly hidden by hair, and the eye is bean- or walnut-shaped, with curved wrinkles around it. The outer line of the trunk is always drawn with a single continuous line, while the inner edge is often hatched to indicate hair. Tusks are often missing altogether (see box p.119).

*An Ice Age artist exploited the shape of a stone plaque for an engraving (left) found at the French rock shelter of La Marche. Like sketches from an artist's notebook, the Gönnersdorf engravings (right) have the freshness of a rapid impression.*

## EARLY HUMANS

The Old Stone Age or Paleolithic period is the earliest and longest phase of human prehistory in the Old World. It began with the first recognizable stone tools (about 2.5 million years ago) and lasted until the end of the Ice Age, about 11,500 years ago. It is divided into three parts: the Lower Paleolithic corresponds to the period of the early human forms, up to and including *Homo erectus*. The Middle Paleolithic is broadly equivalent to the period of the Neanderthals, 200,000 to 35,000 years ago. Debate still rages as to whether Neanderthals interbred with modern humans, or whether they were an evolutionary dead end. They display some sophistication in technology, and their burial sites betray signs of religious belief. They had little

or no figurative art, as far as we know, although growing numbers of simply decorated objects have been found.

Modern humans, *Homo sapiens sapiens*, seem to have originated in Africa more than 100,000 years ago, and spread through Europe between 40,000 and 30,000 years ago, where they coincided with the Upper Paleolithic and the last of the Neanderthals. They were people exactly the same as ourselves in build and appearance, and presumably with the same intelligence. It was they who were responsible for Ice Age art—everything from beads and sculpted figurines to multicolored paintings—and, apart from an occasional mention of Neanderthals, it is the Upper Paleolithic people who are the focus of this chapter.

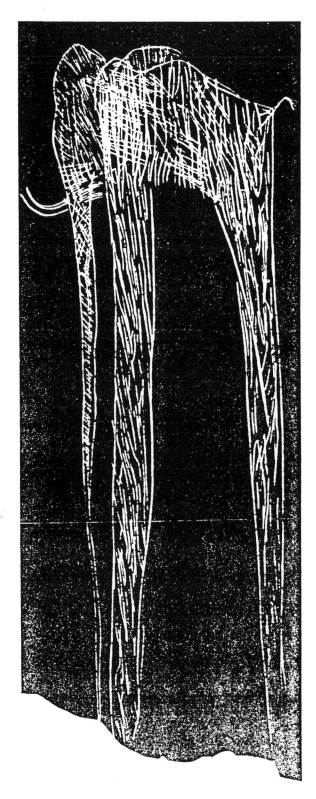

Although often depicted as a simple semicircle at the end of the legs, the feet sometimes have a more realistic flared shape; some even show a protuberance just above the foot, which living elephants have. The short tail curves downward and is sometimes concealed in hair; there is often a tuft of hair, like a brush, hanging down from its end. Overall, the Gönnersdorf mammoths appear detailed and realistic, and each is different. Assuming that groups of fairly similar images on a single plaquette are probably by the same artist, it is reckoned that 12 different artists were responsible for the site's mammoths.

In Ice Age art as a whole, there is usually no correlation between what was depicted and what was hunted or eaten. Gönnersdorf is no exception:

## "Ice Age sculptures and engravings of mammoths have been found in abundance"

the mammoth is the second most frequently depicted animal (after the horse), but mammoth bones rarely feature among the site's animal remains. Reindeer, by contrast, were one of the most hunted species and a major food source, and while reindeer bones abound, reindeer themselves are completely absent from the art.

At Gönnersdorf most of the mammoth engravings were found inside a habitation, near the hearth, with another concentration a few yards away. Depictions of birds in this site were concentrated precisely in those zones devoid of mammoths, while depictions of horses and humans were distributed all over the excavated area. Since the mammoth engravings came from a winter habitation (according to the animal remains) and the birds from a summer one, it is possible that there is a significant link between season and depicted species at this site.

Another major collection of portable mammoths—a set of stone plaques which date to 17,000 years ago—has been found at the French rock shelter of La Marche; they appear to represent only adults, and show no movement. The portable depictions from Gönnersdorf and La Marche together constitute more than half of the 127 known in western Europe.

*The only piece of Ice Age art found north of the Arctic Circle is the extraordinary, elongated mammoth engraved on the tip of a mammoth tusk from Berelekh (drawing above), which is about 12,800 years old.*

*A carving of a mammoth bone (right)—probably a patella (kneecap)—was discovered at Avdeevo in European Russia.*

*A tiny depiction of a mammoth*, as well as the head of an ibex, was spotted on a smooth pebble just 3 in (8 cm) long, found in the Petite Grotte de Bize in the Aude region of France. The site dates to around 20,000 years ago.

Engravings on ivory, similar to that of La Madeleine, have been discovered at Obere Klause in southern Germany, and Mal'ta in south-central Siberia.

The biggest collection of three-dimensional figures comes from Paleolithic sites on the Russian Plain. Among the 40 found so far are ten tiny (1 in by ½ in or 2.7 cm by 1.4 cm) marl carvings from Kostenki. Bigger and more spectacular is the limestone mammoth, 4 in (10.6 cm) long, from Avdeevo, Russia. Ivory statuettes of mammoths were found in the caves of Vogelherd and Geissenklösterle in southwestern Germany; they date to at least 30,000 years ago and are among the oldest known figurative images in the world.

In southern France there are some splendid carvings in antler, forming the hook-ends of spearthrowers (a tool used to give extra leverage to the throw of a spear). The carvings depict birds, fawns, horses, and two particularly fine mammoths—one from Bruniquel and the other from Canecaude.

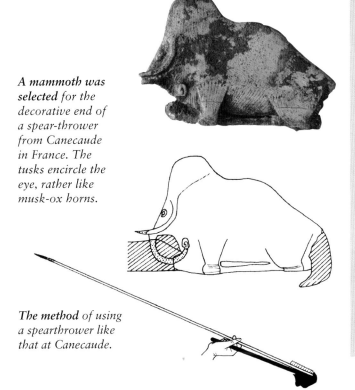

*A mammoth was selected* for the decorative end of a spear-thrower from Canecaude in France. The tusks encircle the eye, rather like musk-ox horns.

*The method* of using a spearthrower like that at Canecaude.

# TUSKLESS MAMMOTHS

Many details in Ice Age depictions of mammoths, such as the small ears and hairy tails in the engravings (above) at Gönnersdorf in Germany, are extremely realistic and coincide with observations on frozen mammoths in Siberia. It is curious, therefore, that although a few of the engravings of the adult mammoths from Gönnersdorf have small, short tusks, most have none at all. Similarly, there are no tusks on the far older mammoth figures from Vogelherd and Geissenklösterle in Germany, nor on an engraving from Kostenki in southern Russia, nor on a number of paintings and drawings on cave walls in France and Spain—for example, in Chauvet cave, only 25 of the (at least) 70 mammoth figures have tusks.

It has, therefore, been suggested that some mammoths had no tusks. An absence of tusks might have been associated with small body size, perhaps through a depletion in natural resources: a mammoth femur found in Gönnersdorf was only 35 in (90 cm) in length, at the low end of the range for woolly mammoths, and meaning the animal stood less than 8 ft (2.5 m) at the shoulder. However, we cannot be sure that this one bone is representative of the Gönnersdorf mammoth population as whole. Two pieces of ivory found at the site were of average size, indicating mammoths with large tusks. Although it is possible that these tusks were already in fossil form when they were introduced into the site by its occupants, two skeletons of similar age to the Gönnersdorf mammoths—found at Condover in England and Praz Rodet in Switzerland—also have average-sized tusks, suggesting that tusks were indeed the norm.

Conceivably, the Gönnersdorf engravings point to a physical difference between the sexes, with the females in some regions perhaps lacking tusks altogether, as in Asian elephants today. Since, however, there is no fossil evidence for this, most specialists would prefer to attribute the lack of tusks in paintings and sculpture simply to artistic licence. At La Marche, for example, some mammoths are depicted with long tusks, some have short tusks, and others have none at all.

# CAVE ART: MAMMOTHS AS DECORATION

It was first suggested in the 1870s that the walls of caves and rock shelters in Europe were decorated with Ice Age art, but the reality of this was not accepted by the archeological establishment until 1902; this was largely by virtue of the discoveries of cave art in France. The debate inspired another wave of enthusiasm that led people to re-examine the walls of known caves, and to explore new ones, in the hope of finding examples of this hitherto unimagined artistic heritage. Decorated caves are still being found, at the rate of about one a year, mainly in southern France and Spain. Most are dated to the Paleolithic period on the basis of the style of their figures, which can be compared with dated specimens of portable art. New advances in pigment analysis and in radiocarbon dating have also been applied to the materials used and have confirmed the Paleolithic date of many caves.

> ## "A single cave in France contains almost one-third of the world's Ice Age depictions of mammoths"

Ice Age wall art encompassed an impressive variety and mastery of techniques. Engraving was by far the most common method used, and the natural formations of walls often directly inspired, or were incorporated in, the representation of figures. In areas of France where the soft limestone could be easily carved, bas-relief sculpture has also been found, always in parts of rock shelters or caves where natural light could reveal the contours of the image. The mammoth of Domme in the Dordogne (see p.124) is an outstanding example of this technique.

Analyses in some caves have also pointed to the use of extenders—i.e. minerals such as talc or potassium feldspar that, when mixed with pigment, make the paint go further and spread more easily when wet; they also produce a darker color than pure red ochre, improve adhesion to the rock, and stop the paint cracking as it dries.

Of the cave sites where mammoths have been depicted, a few stand out from the others. First and foremost is the huge cave of Rouffignac, in the Dordogne (France), with its miles of decorated

*Large mammoth, 5 ft (1.5 m) high, drawn on a semi-vertical wall in the Grande Grotte of Arcy-sur-Cure, France. The tusk departs from the front of the trunk, as at Pech-Merle and Cussac. New figures are discovered in this cave every year as calcite is carefully removed from the ceiling.*

*One of the most remarkable of the 70 mammoth figures in France's Chauvet cave, discovered in 1994, is this example, with its long trunk, straight belly, raised tail, and four tusks seen in frontal view—the black pair are asymmetrical; the engraved pair were added later.*

*Discovered in 2000,* the numerous pristine engravings in the French cave of Cussac, include this remarkable detailed mammoth; the trunk of another can be seen at top center. Note the shagginess, the shape of the feet, and the tusks leaving the front of the trunk, as at Pech-Merle.

*Two mammoths* in France's Chauvet Cave (left), engraved on the wall and both using the same stomach line. They are both about a meter long, and earlier cave bear claw-marks can be seen on the wall beneath them.

galleries, which alone contains 159, almost a third of the depictions of mammoth known in caves—a unique concentration of engravings and black drawings. Some on one low ceiling are so big—more than 6 ft (2 m) in length—that it would have been impossible for the artists to see the whole image at once.

Other notable groups of figures in France are the 70 depictions in Chauvet Cave, the 37 engravings in Les Combarelles (many of them quite cursory, limited to a partial sketched outline), and those of Pech-Merle, Font de Gaume, Bernifal, Jovelle, Chabot, and Arcy-sur-Cure. The mammoth is far more common in the Dordogne and the Ardèche regions than in the Pyrenees and Spain: in the latter region it sometimes seems to have a lack of hair, as at Cougnac, La Baume Latrone, and other French sites, which led to erroneous identifications as warm-climate elephant species such as the straight-tusked elephant *Palaeoloxodon antiquus*. A handful of examples have been found in Spain. There are also fine mammoth depictions of Ice Age date on the walls of a Russian cave in the Urals, Kapova; another Russian cave, Ignatiev, was claimed to contain mammoth depictions, but they are by no means clear, and dating of the cave's pigments now suggests that its decoration belongs to the later, postglacial period.

Most of the mammoths portrayed show few signs of movement, except for little touches of animation such as a raised tail or curled trunk. Nor are the mammoths generally presented in scenes; however, one notable exception is the apparent depiction of a whole group of mammoths in the cave at Rouffignac, drawn as two lines facing each other (see p.100). Pairs of mammoths are sometimes found head to head, perhaps greeting one another.

*A mammoth drawn in black* on a semi-vertical surface of the West Corniche in the Grande Grotte, Arcy-sur-Cure.

*A mammoth found in the cave at Bernifal*, in the Dordogne in France, was painted in yellow on the ceiling of a high natural chimney. Although this has long been a well-known cave, the painting was not spotted until the 1970s.

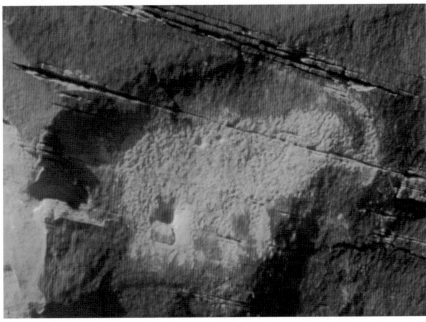

*The rare depictions of mammoths or mastodons* in the southwest of the United States are all petroglyphs, hammered into rock faces. This petroglyph is on the bank of the Colorado River near Moab in Utah.

*Petroglyph of a mammoth* on a rock at Tsagaan Salaa, Mongolia, discovered in the late 1990s. The animal is facing right, with its head at top center, and its trunk tip may be depicted in two different positions.

Occasionally, as in the French cave of Pech-Merle, one finds both detailed depictions and simple abbreviations in the form of characteristic dorsal lines. In several caves, dating to around 30,000 years ago—Pech-Merle, Cussac, and Arcy-sur-Cure (see p.121)—there is a curious stylistic convention, anatomically incorrect, whereby the tusks are drawn from the front of the trunk. One mammoth in Chauvet is drawn with four tusks, one to the right and three to the left.

Outside western Europe and the Urals, no depictions of mammoths have been found in caves or rock shelters. However, a small number of what seem to be mammoth figures have been discovered in the open air on the Colorado Plateau in the southwestern United States. These petroglyphs (figures hammered into the rock) may be the only authentic contemporary representations of mammoths in North America. Whereas all European depictions are of the woolly mammoth, *Mammuthus primigenius*, the Colorado Plateau petroglyphs must be based on the American species, the Columbian mammoth, or conceivably the mastodon.

Most recently, two or three depictions of what seem to be mammoths have been found on rocks in the open-air in the valleys of Tsagaan Salaa and Baga Oigor, in the Altay Mountains of northwest Mongolia.

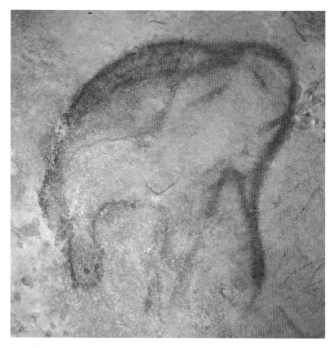

*Small mammoth, 1ft 7in (45 cm) long, drawn in red, retouched with black, on the ceiling of the Salle des Vagues in the Grande Grotte of Arcy-sur-Cure; probably a youngster, judging by the proportions and the lack of tusks.*

*This highly stylized beast drawn with fingers dipped in red clay (right) is one of the most unusual depictions of a mammoth produced in the Ice Age. It was discovered in 1940 in the French cave of La Baume Latrone. The body, 4 ft (1.2 m) long, is in profile, but the short tusks are both depicted as if seen from the front. In contrast to the boldness of the rest of the mammoth, the mouth and trunk tip seem to have been painted with particular attention to detail.*

*In the Grotte du Mammouth—the "mammoth cave"—at Domme in France, a mammoth was sculpted in bas-relief 13 ft (4 m) above the present cave floor. Although it is 4 ft (1.2 m) high, it went unnoticed until 1978, when it was caught in the glare of artificial light.*

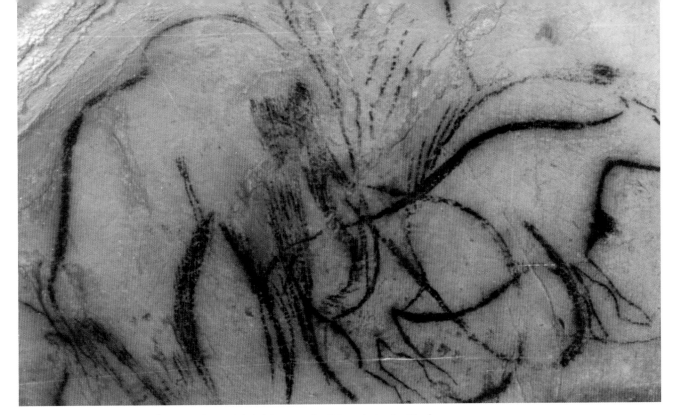

*Mammoths appear among horses* (above right), *bison, and wild cattle in the Black Frieze at Pech-Merle, in the Lot region of France. Experiments with materials similar to those used 18,000 years ago have shown that the entire frieze, consisting of 25 animals and stretching for 23 ft (7 m), may have taken the artist only an hour to complete.*

# HUTS MADE OF MAMMOTH BONES

Some 18,000 years ago the natural accumulations of mammoth bones in eastern Europe provided a convenient source of building materials. Skulls and other large bones were used for the foundations, while roofs probably consisted of hides slung over a structure of wood, or bones and tusks.

*All the bone huts so far discovered are variations on the same architectural theme. The arrangement of the huts, however, varies from site to site: in some cases they may be only a few feet apart, whereas in others they are scattered over a large area.*

*Many of these camps were primarily occupied in the winter and were the scene of a wide variety of activities—cooking, toolmaking, hide preparation, and the production of art objects. How long they were inhabited is difficult to estimate, although the thousands of artifacts at some sites suggest a prolonged stay by a large group.*

Various tools made of **mammoth bone** have been found at the Upper Paleolithic site of Avdeevo in Russia. These include elegant spatulas (see p.130), and two spoons (right).

**Pairs of tusks may have been used** to create imposing hut entrances, or may have been incorporated into the structure of the hut as roof supports. In cold weather, entrances would probably have been kept as small as possible to keep the heat in, as is the case with igloos.

**Scores of mammoth carcasses** were required to produce the number of bones present at each site. Some huts contain bones from a number of mammoths, which had clearly died over a span of several thousand years.

# MAMMOTH-BONE HUTS

During the last Ice Age quantities of mammoth bones lay strewn across various parts of central and eastern Europe—enough for humans to consider using them to build dwellings in sites where caves and rock shelters were not available. A total of more than 70 mammoth-bone dwellings is now known from about 15 sites on the central Russian Plain, particularly in the Ukraine along the River Dnepr and its tributaries, with single examples also in southern Poland (Kraków, Spadzista Street) and Moravia (Milovice) in the Czech Republic.

As long as 40,000 years ago, Neanderthals had used mammoth bones for the construction of shelters. At the Ukrainian site of Molodova, a 26 x 16 ft (8 x 5 m) oval area of mammoth bones containing 15 hearths is thought to have been a structure in which the bones held down skins stretched over a wooden framework. But the most spectacular structures of this type were produced by fully modern people between 35,000 and 17,000 years ago.

At first sight the remains look like disorderly heaps of bones—in fact mammoth-bone dwellings were not recognized as such by archeologists for many years. Since medieval times, the villagers of Kostenki in Russia had been finding large bones (the name comes from *kost'*, the Russian word for bone); a legend claimed that they were the bones of antediluvian giants. In the 18th century, Tsar Peter the Great took an interest in the finds, but attributed them to war elephants from a wandering army of ancient Greeks.

A century later Kostenki was recognized as a prehistoric occupation site, but the existence of mammoth-bone structures was not suspected until excavations during the 1920s: until then the bone heaps were thought to be food refuse.

The huts were round or oval, between 13 and 22 ft (4 and 7 m) across at the base, enclosing from 86 to 258 sq ft (8 to 24 m²) of living space. At some sites they are set out in rows, at others in a rough circle or rectangle, depending on the terrain: at Yudinovo, 250 m (400 km) SW of Moscow, two of the four huts are only one metre apart, as are the three circles at Kraków, whereas at Dobranichevka in Ukraine (right) the four circles

*Even after 18,000 years of disuse, mammoth-bone hut No. 4 at Mezhirich, in the Ukraine, gives a powerful impression of its original scale. The structure has collapsed inward, producing the jumbled mass of bones in the living area. By sectioning off the site with ropes and carefully numbering the bones, archeologists were able to reassemble the site like a jigsaw puzzle.*

of upright mammoth bones are scattered over several hundred square metres and are never less than 65 ft (20 m) apart.

Mammoth skulls, jaws, shoulder blades, and other large bones formed the foundations, and tusks may have been used to create a framework for a porch or entrance; in some cases they were joined together with a bone sleeve, while in others they were simply left in their sockets in the skull. The roof is likely to have been a wooden frame covered with animal skins or turf held in place by more bones and tusks. In some, the floor was 16 in (40 cm) below surface level, making the huts semi-subterranean.

Many of the dwellings include the site of a fireplace in which the occupants burned bone as fuel, no doubt because timber was so scarce in these regions. Bone burns well as long as it retains its collagen and fat content; indeed, some Eskimo groups today still use bone as a fuel source. The first few minutes are smoky and smelly, until the fat has been scorched off, but after this bone provides a long burn, with a steady flame and a high heat yield. Suitably fresh or "green" bone was probably stored in pits in the permafrost (see p.155) or simply scavenged from carcasses preserved by the extreme cold.

The best-known bone-hut site is probably Mezhirich in the Ukraine, where at least five dwellings, constituting a winter camp, have been found dating to about 18,500–18,000 years ago. The construction of dwelling No. 1 is typical; in this, a total of 25 skulls was placed in a semicircle to form the interior base wall, their frontal bones facing inward and their tusk sockets buried in the ground. These were supplemented by 20 mammoth pelvises and 10 limb bones, also fixed into the ground. On top of this foundation were 12 more skulls, 30

*Remains* of one of the mammoth bone huts at Dobranichevka.

*Excavation continues today* at sites with remains of mammoth bone huts: these bones were exposed at Gontsy in 2006.

shoulder blades, 20 femurs, 15 pelvises, and segments of 7 vertebral columns. Higher still, and probably used to hold down the roof-hides, were 35 tusks.

Some 95 lower jaws, placed chin-down, one on top of the other in a herringbone design, formed an outer wall, which has various interpretations. It may have acted as a retaining wall or held down the hides that covered the shelter; alternatively, it may have provided a layer of heat insulation between the hides and the snow that drifted against the jaw bones—or may simply have been a supply of raw material and fuel. Mammoth leg

## "The remains of more than 70 Ice Age huts built out of mammoth bones have been found across the Russian Plain"

bones were placed upright at the small entrance. Inside was a mammoth skull decorated with zigzags and dots of red ochre, which one Soviet archeologist interpreted as representing flames and sparks.

Other dwellings at Mezhirich are variations on this basic theme, with jaws placed chin-upward, for instance, or with a range of different bone types used in the base wall. Some bones have holes drilled into them and may have been held in position by bone or wood inserted into them as pegs; alternatively, they could have been used for suspending garments to dry, or for hanging up meat and other necessities, or even as peepholes in the outer layer of the walls or roof. At Mezin hut 1, the crests on 52 of the 53 shoulder blades had been worn down or snapped off, so they could be stacked more compactly. Features

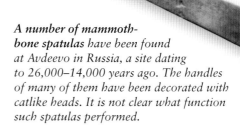

of this kind demonstrate vividly the tremendous ingenuity of Ice Age people in adapting available materials to their needs.

The amount of labor and materials involved in these structures varied. Dwelling No. 1 at Mezhirich, for example, is reckoned to contain some 385 bones weighing a total of 46,300 lb (21,000 kg); other huts at the site contained between 33,000 and 42,000 lb (15,000 and 19,000 kg) of bone. At other sites no more than 2,200 lb (1,000 kg) were used for each dwelling. Archeologists have estimated that it would have taken 10 people at least five or six days to build dwelling No. 1, and four or five days for each of the others. The smaller dwellings at other sites may have required no more than half a day each. To this must be added the labor involved in collecting the bones in the first instance: some idea of what this entailed can be gauged from the fact that a defleshed and dried mammoth skull with relatively small tusks weighs a minimum of 220 lb (100 kg), and often far more, while the other large bones are by no means light. However, since the huts

*A number of mammoth-bone spatulas* have been found at Avdeevo in Russia, a site dating to 26,000–14,000 years ago. *The handles of many of them have been decorated with catlike heads. It is not clear what function such spatulas performed.*

## "The age of the mammoth bones used in individual huts spans 10,000 years"

were no doubt very sturdy and robust and would last from one season's visit to the next, it was probably worth investing a considerable amount of time and effort in them.

It is obvious that a huge number of mammoth skeletons was available in these areas. Some of the bones used in the dwellings might conceivably have come from kills or even mass drives by the occupants, but it is more probable that the building materials came from carcasses, the product of natural deaths or of the predation of mammoths by other animals. Rivers and streams would also have created natural accumulations of bones, and these would have remained unburied and relatively fresh under periglacial conditions. (The question of whether mammoths were hunted, or whether their remains were simply scavenged, is discussed more fully in Chapter 5.)

The inconsistent state of preservation of the bones suggests that some were recovered from long-dead skeletons: radiocarbon dating has revealed that bones used within a single structure at Mezin, in the Ukraine, range in antiquity from 27,000 to 17,000 years. Many bones had been gnawed by carnivores: studies of modern elephant carcasses left in exposed sites show how scavengers soon scatter the bones as they pull carcasses apart.

The Mezhirich site contains the remains of at least 149 individual mammoths, and other Ukrainian sites also have about 100 each. Some of the bones might have been accumulated gradually, over the years, but in the case of Mezhirich dwelling No. 1 the 25 skulls used to build its foundation had to be collected before construction could begin. If the greater part of the material came from natural mammoth "cemeteries" or accumulations of bones, then it must be assumed that the occupation sites were placed fairly close-by to avoid the enormous effort of transporting the bones any distance.

The number of people occupying these hut groups is impossible to estimate, since it is not clear whether all the structures were dwellings, and whether they were all built and used simultaneously or in sequence. The duration of occupation is equally elusive, though the thousands of artifacts at some sites suggest a prolonged stay by a large group. Archeologists working at Mezin have assumed that the five huts formed a settlement for a number of families (about 50 people). Modern populations in the frozen north, whose members still construct huts of similar shape (though no longer of bones), use them for 15 to 20 years, as long as the wooden frame is sound; then they build a new settlement elsewhere. This may also have been the pattern with the Ice Age huts, although the useful life of a mammoth-bone hut may have been considerably

MAMMOTHS AND HUMAN CULTURE

# THE MAMMOTH-BONE ORCHESTRA

Northern peoples today, such as the Eskimo and Samoyed, always have a dance hut in their villages. This is the same size as a dwelling, but is set aside for song and dance, entertainment and ritual, and is used especially in the festivals of fall and winter, at the start of the winter hunt. The site of Mezin in the Ukraine, dating to about 24,000 years ago, contains a mammoth-bone hut which may have been devoted to a similar use. Although the interior space was probably too constricted by roof supports to allow for dancing, the floor appears to have been kept clear of domestic refuse—a sign that it was perhaps used for ritual purposes.

A set of mammoth bones painted with red ochre and a reindeer antler hammer were found in a group on the floor of the hut. Originally they were thought to be art or cult objects. However, subsequent analysis revealed areas of surface damage on all of them, as well as signs of smoothing, thinning, polishing, or rubbing. The conclusion must be that they formed a set with some common function that involved concentrated blows, as well as rubbing and polishing through prolonged contact with hands and perhaps fur clothing. The reindeer antler has the polish of long use on its handle, while its working surface is very worn and its spongy pores have reddish ochre in them, picked up from the decorated surfaces on the mammoth bones.

A controversial interpretation of this evidence is that these bones were used as musical instruments, in a kind of Ice Age orchestra. The bones were presumably percussion instruments, struck with hammers. A mammoth shoulder blade, decorated with linear and zigzag stripes in red ochre, bears traces of polish on the neck, corresponding to the positions of the palm and thumb, suggesting that this was where it was held with the left hand, while the right hand struck the body of the instrument with a hammer. Other areas of wear possibly indicate that the bone was hit in different places to vary the tone of the note produced. A femur had had the soft, spongy material extracted from

*Mammoth-ivory flute discovered in Geissenklösterle, Germany.*

inside, perhaps to increase its resonance. It may have been played horizontally, like a xylophone, perhaps on a support rather than the ground, judging from other polished areas. A half-pelvis and two jaw bones, likewise painted with red parallel stripes, appear to have been used in a similar way. The two jawbones were placed on their undecorated left sides, and the right sides—both painted with red stripes—were struck. The teeth were removed from one of them, perhaps to alter its resonance. A piece of bone from a skull, decorated with spots of red color, was probably used as a drum: the cranium's cellular structure creates an unusual resonance, as experiments with modern elephant skulls have shown.

Finally, among this set of bones was a "bracelet" made of five very springy rings of ivory, perforated so that they could be tied together, and all incised with a herringbone design. This too may have been a kind of musical instrument, producing a sound similar to castanets. The hut also contained 2 or 3 kg of red and yellow ochre, in four concentrations, as well as bone pendants, needles, and awls: some scholars have suggested that these may have been used to prepare for theatrical performances.

After restoration and conservation in the Hermitage Museum, all the original Mezin instruments were played with bone hammers by a collection of percussionists under the direction of V. I. Kolokolnikov of the Kiev State Philharmonic. The musicians were able to produce a variety of resonant sounds: using their knowledge of the music of northern peoples today, they were able to give a rendering of what Ice Age music might have sounded like.

In 2004 a flute carved from mammoth ivory was unearthed in 31 fragments from the South-West German cave of Geissenklösterle. It is 7 in (18.7 cm) long, has three fingerholes, and is more than 31,000 years old. So it is now clear that mammoths were indeed used to make music in ancient times.

longer, since bone is far more resistant than wood to frost, humidity, and insects.

Most of the mammoth-bone hut groups have large storage pits, 6–26 ft (2–8 m) long, which were dug up to 5 ft (1.5 m) deep into the permafrost and then filled with hundreds of mammoth and other bones, presumably as a supply of food, fuel and construction material. Most also have distinct "activity" areas, where various stone or bone tools were manufactured. For example, at Gontsy, a Ukrainian site where excavation has resumed in recent years (see p.129), one dwelling area is surrounded by ashy areas with charcoal, bone, and stone tool fragments, while another hut has a hearth for ochre preparation, and a third has butchering areas.

## MAMMOTH-BONE TOOLS

A wide range of tools and even furniture was also made out of mammoth bones. At Kostenki the shoulder blade of an adult mammoth seems to have been set vertically in the ground, inside a hut; its upper surface, 13–16 in (33–40 cm) above the ground, would have reached the chest of a seated person, and the dents and notches in it suggest that it may have been used as some kind of work surface, anvil or small table. Mammoth foot bones bearing percussion marks may also have been used as anvils, even by Neanderthals—as, for example, in caves in the Crimea such as Kiik Koba. These compact, cuboidal bones, about 7½ in (20 cm) across, would have been ideal for this purpose.

A mammoth rib from Kostenki appears to have been used as a palette for mixing colors. Shoulder blades were also used in funeral rites, being placed over a number of Upper Paleolithic graves: for example, over a grave

*A cleaver made from a mammoth shoulder blade was found at the Lange-Ferguson site in South Dakota where mammoths had been butchered. It was sharpened by flaking away pieces of bone.*

in a deep pit at Kostenki. The practice seems to have been particularly prevalent in Moravia where the Brno II burial lay beneath a scapula, while a female grave at Dolní Věstonice was covered by a scapula with part of a mammoth pelvis on top of it. In 2005, a burial of two infants, of about 30,000 years ago, was unearthed at Krems-Wachtberg (Austria)—they too were covered with a mammoth scapula.

> ## "Cleavers, spatulas, wrenches, and shovels were fashioned out of mammoth bone"

The most impressive example was the collective burial excavated at Předmostí in 1894, where 8 adults and 12 youngsters lay in an area 13 by 8 ft (4 by 2.5 m), covered by numerous large bones including two shoulder blades. Were these large, flat bones simply convenient covers, or do they denote some belief in the mighty mammoth protecting loved ones in the afterlife? The animal may also have protected the living: at some Russian sites, tail vertebrae and some foot bones, still in anatomical order, have been found specially placed (or intentionally hidden) in small pits near the hut walls.

A wide variety of tools was made in different parts of the world by modifying individual mammoth bones. In North America, for example, there are sites such as Tocuila, Mexico (see p.73), revealing the deliberate quarrying of mammoth bones around 13,000 years ago for the production of cores and flakes. In the United

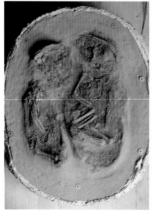

*In 2005, the burial of two infants, of about 30,000 years ago, was unearthed at Krems-Wachtberg, Austria—they were covered with a mammoth scapula.*

States, at the Lange-Ferguson site in South Dakota, an adult and a juvenile mammoth 12,800 years old were found with two heavy cleaver-choppers made from the flat part of a mammoth shoulder blade, and study under the microscope confirms that they were employed in butchering the carcasses. A cutting tool ("flake knife") was also made from a segment of mammoth long bone; it was probably used to sever the fibrous tissue around the spine, since it was found tightly wedged vertically against some chest vertebrae.

At Murray Springs, Arizona, an intriguing tool was discovered in 1967. It was a perforated baton, 10 in (26 cm) long, made of a mammoth femur and dating to about 13,100 years ago. It is thought to be a shaft-wrench—a tool for straightening wood or bone to make the shafts of spears, similar to tools used by Eskimos. Shaft-straighteners have also been found at Mezin (in ivory) and Molodova, and a square-headed specimen was found in the Sungir graves (see p.134).

In the Old World, tools of mammoth bone were already in use by Neanderthals—notably a handaxe from Rhede (Germany), and a number of ribs and fibulae from Salzgitter-Lebenstedt (Germany), which were pointed and/or flattened. The latter site also yielded a short, notched triangular point made from a mammoth longbone (see p.152).

The later sites of Moravia and Russia have yielded a wealth of mammoth-bone tools. At Předmostí and Kostenki there are paddle-shaped shovels, cut from long-bones, which are 10 in (26 cm) long with blades 2 in (6 cm) wide, and there are bone handles of axes and picks, with oblique perforations. Předmostí has what appears to be a fishing hook of mammoth bone, dating to around 30,000 years ago, and there are ribs, sharpened at one end by chopping, which could be digging sticks.

Although fresh bone is more plastic and amenable to flaking and working, under the periglacial conditions of the Ice Age the raw material would have remained fresh and workable for long periods, and could therefore have been collected from long-dead carcasses. However, bone is by no means the only such material to be obtained from mammoths: they also provide huge quantities of ivory.

*The oldest known "art object" made from mammoth, this segment of a molar was polished—and one of its faces covered in red ochre—by Neanderthal hands 100,000 years ago. It was found at Tata, near Budapest.*

## OBTAINING IVORY

Consider the daunting task facing an Ice Age sculptor, armed with nothing more than the simplest stone tools, attempting to transform an enormous mammoth tusk into delicate figurines, bracelets and beads. Tusks could be up to 10–13 ft (3–4 m) long and weigh 185 lb (84 kg); and ivory itself is hard and unyielding.

At times it was easy to obtain the tusks: some mammoth skulls at the site of Pushkari I have empty but undamaged sockets, which shows that at least some of the 65 tusks at the site were simply collected after they had become dislodged naturally. However, to remove the tusks from a freshly killed animal, or from a scavenged carcass, would have involved loosening them from their sockets by blows with large stones. Some specimens from sites such as Kostenki I bear the traces of cracks and splintering from hard blows which shattered their outer layer of ivory: it is thought that large flint axes, up to 18 lb (8 kg) in weight, were used for the purpose.

> "Workable pieces of ivory had to be cut from the huge, tough tusks with Stone Age tools"

There were several ways of cutting through a tusk. The most direct was, quite simply, to chop through it with stone axes. Alternatively, since it is easier to chisel ivory than to chop it, a deep, circular groove might be chiselled around the cylinder. When only a narrow neck of ivory remained, it could be broken by a sharp blow, perhaps simply by striking the tusk against a rock.

A tusk could also be split along its length. The crudest method was to strike off irregularly shaped flakes by hitting the tusk with a pointed stone tool. Where more regular shapes were desired, longitudinal grooves could be cut with a sharp-edged flint tool called a burin before the flake was struck off.

Obtaining long, thin pieces of ivory was more difficult. Craftsmen in Europe could apply techniques learnt through antler-working, which was widely practiced in the last Ice Age, but the process was more complicated for ivory. It was relatively easy to cut grooves down an antler and then prise strips out, away from the soft, spongy center, but since ivory does not have a soft center, similar strips had to be chiseled out. After making the longitudinal grooves, the craftsman had to split off the strip, probably by using a bone chisel in the same way that modern Eskimos work walrus ivory. One remarkable unfinished specimen from the Russian site of Eliseevich has preliminary shallow grooves cut down the whole length of a tusk.

## FROM TUSKS TO ART OBJECTS

Having broken down the raw material into pieces or strips of ivory, the Ice Age artists then set about fashioning it into an impressive range of tools, figurines, and ornaments. They used a variety of techniques, involving whittling, cutting, and engraving.

At Mezhirich and other sites, some huts contained areas where ivory seems to have been worked extensively. Mattocks—chisel-like tools—were produced from sections of tusk and beveled at one end to make a massive kind of spatula that may have been used to dig pits in the hard ground. Narrow ivory flakes were also sharpened into stabbing or thrusting weapons.

At Eliseevich a 10-in (26-cm) dagger was made from the end of a tusk, its natural point sharpened by whittling. The handles of such tools were often engraved with cross-hatched cuts to provide a firm grip.

The Moravian sites contain a wealth of ivory tools, such as a spoon from Dolní Věstonice

*The body of an old man* in a 30,000-year-old grave at Sungir in Russia lies adorned with ivory bracelets and thousands of beads.

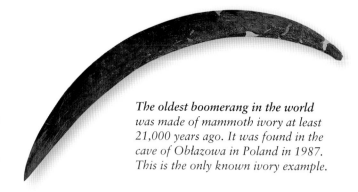

*The oldest boomerang in the world* was made of mammoth ivory at least 21,000 years ago. It was found in the cave of Obłazowa in Poland in 1987. This is the only known ivory example.

decorated with an engraved geometric motif; while the German site of Kniegrotte produced a kind of large comb as well as a fragment of a unique ivory harpoon, with barbs on both sides, and with engraved decoration along its length. At Mezin, eyed needles made of ivory have been recovered.

Ice Age craftsmen were also able to bend or, alternatively, straighten ivory. A child's grave at Mal'ta in central Siberia contained a diadem made with a thin hoop of ivory. Fresh ivory could not have been bent into a hoop; dry ivory must have been forced into the necessary curvature. Conversely, the Ice Age children's graves at Sungir in Russia, dating to around 30,000 years ago, contained two heavy spears of straightened ivory 8 ft and 5 ft (2.4 m and 1.6 m) long. How these were made remains a puzzle, considering the marked natural curvature of mammoth tusks. Steaming is the most likely solution: the ivory was probably soaked in

## "The bodies of children buried at Sungir were each adorned with about 3,500 beads made of mammoth ivory"

water for a long time before being held over a fire and heated to 250°F (120°C) or more in damp conditions to prevent it from cracking when its shape was altered. In an interesting experiment carried out some 50 years ago, the Russian scientist Mikhail Gerassimov thoroughly soaked a piece of ivory for five days, then wrapped it in a pre-soaked fresh animal skin and placed it in the hot ashes of a camp fire. After one and three-quarter hours the soft skin had charred and fallen to pieces, while the ivory was extremely hot. On cooling it could easily be whittled with a flint knife, while thin strips were flexible enough to bend. Siberian and Alaskan Eskimos have an alternative method of softening bone and ivory for carving: they soak them in urine.

Another object that may have been made in the same way to achieve the required curvature is the ivory boomerang found in the Polish cave of Obłazowa in 1987. The oldest boomerang in the world, dating to about 21,000 years ago (though bones from the same layer have been dated to more than 30,000), it has a span of 70 cm, is up to 6 cm wide and up to 1.5 cm thick. One side preserves the external, convex tusk-surface, whereas the other has been polished almost flat; both ends are flattened, and overall the boomerang has an elegant aerofoil section with thin,

## "Many of the most striking carvings of prehistory were made from mammoth ivory"

tapering edges attesting to the great skill of the carver. Experiments with a cast suggest that it was almost certainly not designed to return to the thrower but was of the type used by modern Aborigines simply to stun or kill prey animals. Curved killing sticks of this kind are by no means restricted to Australia, since wooden specimens are known from the prehistory of Holland and Jutland and even from Egyptian tombs. However, the very existence of this unique ivory object is of great importance for our knowledge of Ice Age hunting skills, as killing sticks can be accurate up to about 650 ft (200 m), much farther than a man can throw a spear or stone. The Sungir burials included numerous ivory objects besides the remarkable ivory spears.

There were three bodies: a 60-year-old man in one grave, two children aged 8 and 13 placed head to head in the other. They were surrounded by ivory staves, daggers, long bodkins, small carvings (including a flat, perforated ivory horse with traces of red coloring on it) and two pierced ivory discs. Around the man's upper arms were

*Ivory figure from the SW German cave of Hohlenstein-Stadel, which is half-feline and half-human, and more than 30,000 years old.*

ivory bracelets, with holes for lacing them together (like the bracelet of Mezin, see p.136). Each body was adorned with about 3,500 beads of mammoth ivory, arranged in rows across the forehead, across the body, down the arms and legs, and around the ankles. It is probable that these beads were strung on lengths of sinew, which were then attached to items of clothing that have since disintegrated.

It has been estimated that it would have taken about 45 minutes to make each Sungir bead, if the whole process of cutting the tusk, shaping the bead and drilling the hole is included. This means that each body had 2,625 hours of beadwork buried with it. The standardized and uniform appearance of these objects

Many prehistoric objects carved from mammoth ivory show great craftsmanship and a remarkably sophisticated sense of design. A herringbone pattern has been used in **the broad bracelet from Mezin (1)**, which is about 18,000 years old, while the **ivory plate (3) from Mal'ta in Siberia** is decorated with spiralling dots, which may represent a kind of calendar. **Ivory bird (2)** only 2 ins (4.7 cm) long, found in 2002 in the SW German cave of Hohle Fels, and more than 30,000 years old. The **tiny pendant from Dolní Věstonice (4)** dates to about 26,000 years ago and is only 1 in (2.5 cm) high. One of a series of eight, it may represent stylized breasts or testicles.

*The **thumb-sized ivory head** found at Dolní Věstonice in Moravia dates to around 24,400 years ago. The elaborate coiffure has led to the assumption that it is female, but this may not be the case.*

*The gaze of a woman from 27,000 years ago is captured in mammoth ivory on a small nude figurine, standing 3 in (8 cm) tall, which was found inside a dwelling at Mal'ta. Siberian figurines are unusual in their detailed depiction of hair and facial features.*

suggests that they were produced by a limited number of people.

In western Europe, far more ivory beads have been found in places where people lived than at burial sites. A series of beads in various stages of manufacture has been unearthed in the southern French rock shelter of Blanchard, dating to about 30,000 years ago. These demonstrate clearly the sequence of production. Small rods of ivory, up to 4 in (10 cm) long, were circum-incised and then snapped into sections, separated into pairs, then worked in a dumbbell form before being perforated and separated for the final shaping and polishing.

Given the crude nature of most Stone Age tools, the production of ivory tools, weapons and beads is impressive enough. But it is the statuettes and figurines —and notably the "Venus" figurines—produced by prehistoric carvers that show that they were capable not simply of remarkable feats of craftsmanship, but of startling artistry as well.

Ivory is easy enough to engrave along the grain, but not across it. Nevertheless, by scraping away the surface ivory it was possible to make a design stand out from a background, rather like a cameo, and images of figures were produced in this way, as in the Aurignacian human figure with raised arms from Geissenklösterle, Germany.

Fully three-dimensional figurines were also carved from ivory at an early date: indeed, some of the earliest known pieces of Ice Age art are the ivory statuettes which come from a series of sites in southwestern Germany. These include animal and human figurines from Vogelherd (see p.7) and those of Geissenklösterle (fragments of two mammoths and a feline). The astonishing male statuette from Hohlenstein-Stadel, which dates to around 30,000 years ago, suggests strongly that it must have been preceded by a long tradition in carving ivory.

The rich collection from France includes a delicate, tiny human head from Brassempouy; the Venus figurine from Lespugue, with her arms resting on her monumental breasts; the superb horse statuette from Lourdes, and the two reindeer from Bruniquel. In Moravia, another rich collection includes the well-known "female" head from Dolní Věstonice, and the mammoth from Předmostí. Russia and Siberia have yielded a vast collection, including many female statuettes and stylized birds, several of which are decorated with elaborate geometrical markings. In 2001, a wonderful bison figure, dating to around 26,000 years ago, was unearthed in a pit at the open-air site of Zaraisk, near Moscow. It is 4 in (10 cm) high and 6 in (16 cm) long, and bears traces of red and black pigments. Many other carvings likewise still bear traces of red coloring.

Some idea of how the Russian figurines were made is provided by the finds at Kostenki and other sites. These are so numerous that there are examples of every stage of creation, from crude rough-outs to highly polished finished items. One piece of ivory from Gagarino—a site which has yielded eight female figurines in ivory—has been carved to form two human figures of different lengths, attached at the head like Siamese twins. It probably represents two figurines that have not yet been separated.

*Ivory figure of a bison,* unearthed at Zaraisk, European Russia in 2001; it is 4 in (10 cm) high.

> ## "There may be 10 million mammoth carcasses lying in the deep freeze of the Siberian permafrost"

Experiments have been carried out in carving statuettes in ivory: to make a figure, a piece of tusk was probably first pried out by cutting two deep grooves into the ivory; this was rubbed into rough shape with sandstone, and a burin or other form of sharp blade was used to carve the legs and other details. The finished piece may then have been polished with leather straps—although long handling would also have smoothed away some of the inevitable marks of the

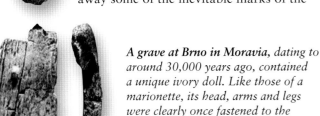

***A grave at Brno in Moravia,*** *dating to around 30,000 years ago, contained a unique ivory doll. Like those of a marionette, its head, arms and legs were clearly once fastened to the body. The doll is 5 in (13 cm) tall.*

arduous and time-consuming labor that ivory carving demanded. Carving ivory in this way takes many hours, even days, of effort.

## MAMMOTH IVORY IN THE MODERN WORLD

The importance of the mammoth to prehistoric people in some parts of central and eastern Europe can hardly be overestimated: in fact, human occupation of some areas seems to have depended almost entirely on the existence and the presence of the animal. Its bones were used for construction and as fuel, as well as for tools and even musical instruments, while its tusks were fashioned into ivory tools, jewelry, art, and ritual objects.

Mammoth ivory continued to be exploited even after the mammoths' extinction and, thousands of years later, it is still being used. Even today it remains a vital resource to some communities of craftsmen.

Fossil ivory from Siberia, called *mamontova kost'* ("bone of the mammoth"), has been known since very ancient times, and is mentioned in Chinese writings of the fourth century B.C.—it was probably already being imported to China at that time. There are records of it being exported from Russia to central Asia and Europe in the 10th century A.D: the throne of the Mongol Khan Kuyuk (1206–48), grandson of Genghis Khan, was allegedly made of it.

The first fossil ivory known to reach western Europe was a tusk purchased from Samoyeds in Siberia and brought to London in 1611. After Russia conquered Siberia in 1582, it became a regular commodity, and the tsars tried to monopolize this trade. From the mid-18th century onward, the Siberian ivory industry expanded greatly, and the statistics are impressive. For instance, a single collector returned from the New Siberian Islands in 1821 with 18,000 lb (8,165 kg) of ivory, equivalent to 50 animals. During the first half of the 19th century, 36,000 lb (16,330 kg) were sold at Yakutsk every year in addition to the smaller amounts sold at other towns. Some authors speak of large boats on the Lena River laden with mammoth ivory, and it has been estimated that during the first 250 years after the Russian occupation tusks from at least 46,750 animals must have been excavated—and that in later years an average of 250 animals needed to be found annually to meet the demand. In the early 19th century, mammoth ivory became a significant source of raw material for the manufacture of billiard balls, piano keys, and ornamental boxes. In 2004, craftsmen in Scotland began using mammoth ivory as decorative inlay on bagpipe drones.

Siberia was once considered to hold an inexhaustible supply of this "white gold;" but it is a non-renewable resource which needs to be protected. Locals, however, claim that they have industrial quantities to offer—some estimates suggest that there are 10 million mammoths still lying in the Siberian deep freeze.

## MAMMOTHS TO THE RESCUE?

In recent years mammoth ivory has come to further prominence, as a result of the ban on international trade in elephant ivory. If there is so much mammoth ivory in the world, could it be used to help save its modern counterparts from extermination? Certainly it has proved its worth to many contemporary ivory carvers.For example, the little town of Erbach im Odenwald in Hessen, Germany, has specialized in ivory carving since 1783, when the "Ivory Count," Franz I, introduced the art to the community and opened a workshop for that purpose in his palace for the bone carvers and wood turners of the area to improve their economic lot. Ever since, Erbach has been the center of ivory carving in Germany, winning renown for its carved trinkets, bracelets, boxes, and seals.

*A continuous tradition of mammoth-ivory carving links the prehistoric with the modern. This 19th-century casket is typical of the vast range of ivory products made in Yakutia, where the craft survives to this day.*

## MODERN MAMMOTH-IVORY OBJECTS

The art of mammoth-ivory carving has a long tradition in Russia—both Peter the Great and Catherine the Great pursued it as a hobby. It is witnessing a renaissance in modern Russia, where figurines and jewellery make popular tourist souvenirs.

The legislation of June 1989 following the Lausanne Conference prohibited the purchase of all new elephant tusks and threatened to put an end to the community's livelihood: the 30 remaining Erbach carvers, who once numbered 1,500, still used 1.5 tons of elephant ivory—between 200 and 300 tusks—every year. Then the town's mayor remembered the stocks of mammoth ivory in Siberia: being extinct, mammoths are not an endangered species. The first consignment of more than a ton, from Yakutia, arrived by truck in 1990. In 2006 there were still 10 ivory workshops active in the area, using both mammoth ivory and the seed of the ivory nut palm, often called "vegetable ivory."

Meanwhile, a considerable output of mammoth-ivory carving, both for the local tourist market and for export, continues to flow out of Yakutia. A similar craft industry has developed in Alaska.

There is a slight difference in color between mammoth and elephant ivory. Mammoth ivory is usually not pure white but creamy or brown, and there is a slightly coarser texture. Some of the Japanese ivory craftsmen—in 1988 they imported 106 tons of African tusks for piano keys and personal seals—have expressed disappointment in mammoth ivory, complaining of poor quality and cracks. On the other hand, exquisite and intricately carved objects have been made from mammoth tusk throughout the centuries in Russia, and are a match for anything made out of elephant ivory.

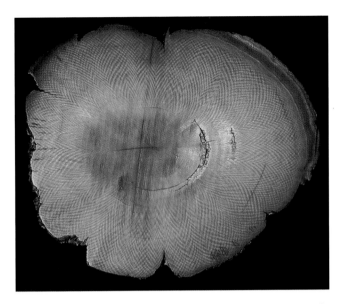

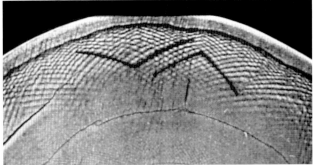

*Elephant ivory (above); mammoth ivory (below).*

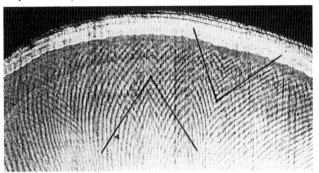

*In cross-section, mammoth and elephant tusks show a characteristic criss-cross pattern. They differ, however, in the angle between the diagonals, helping to separate mammoth from illegal elephant ivory, at least in the unworked state.*

With the growing global interest in mammoth ivory, the price has gone up. In 2005, at an auction of tusks in Yakutia (where between 30 and 50 tons of tusks are gathered every year during the summer thaw), one 33 lb (15 kg) tusk fetched around $2,500.

Although, in theory, the use of mammoth ivory might reduce the need for tusks from the endangered living elephants, the situation is more complex. Under current legislation, trade in mammoth ivory is legal, while that in elephant ivory, with few exceptions, is not. Customs officials are therefore faced with the difficult task of separating the two. It is relatively simple to distinguish ivory from bone or walrus tusk, since it alone has a cross-hatched grain. It can, however, be difficult to tell pale mammoth ivory from that of modern elephants, especially once carved. Proper analysis (for example, by microscopic observation and radiocarbon dating) is expensive and time-consuming, but if shipments are allowed to pass untested, what purports to be mammoth ivory, but is actually poached from elephants, may slip through the net, and there is clear evidence that unscrupulous dealers are labeling elephant ivory as mammoth for just this reason.

Thus, far from saving its modern cousin, the mammoth may unwittingly provide a means for its continued persecution by prolonging the ivory trade and masking an illicit trade in elephant ivory. At the same time, the demand for mammoth ivory fuels an ever greater market for excavation of tusks in Siberia. While some excavators respect the scientific worth of the finds (such as attached skulls or even carcasses) associated with the precious tusks, and donate or sell them to museums, others, rather like the poachers of living elephants, take the tusks and leave the rest to decay. Thus the precious remains of an extinct animal are lost to science.

*A **shipment** of mammoth ivory seized by customs guards at an international port.*

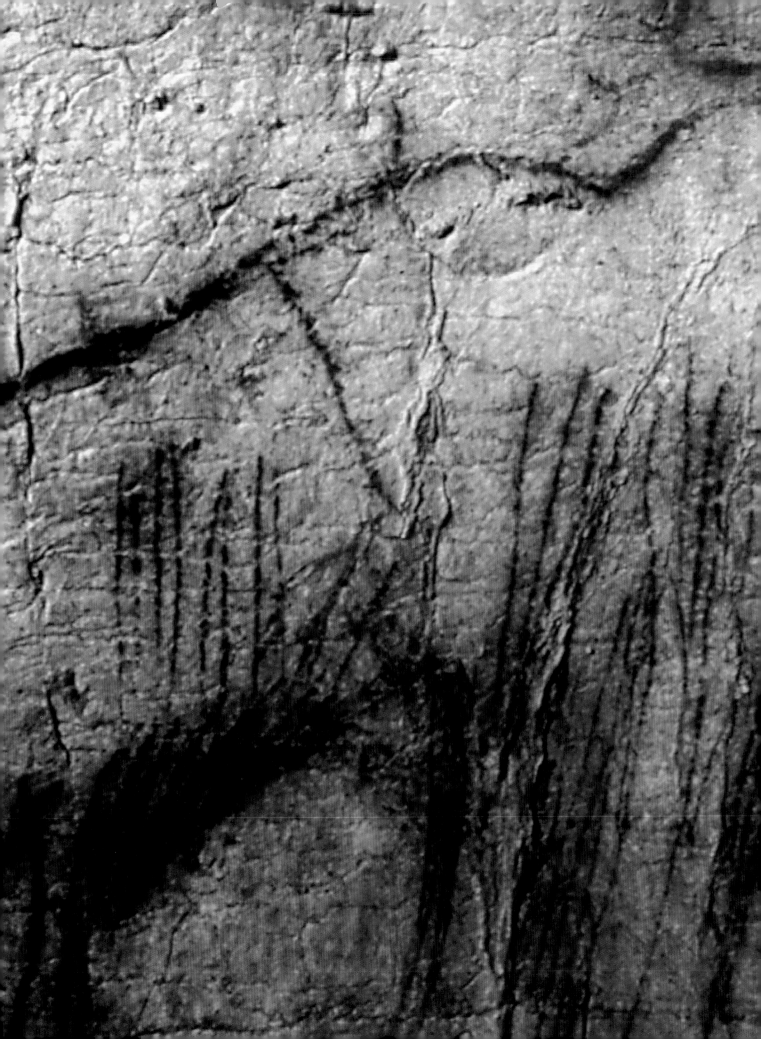

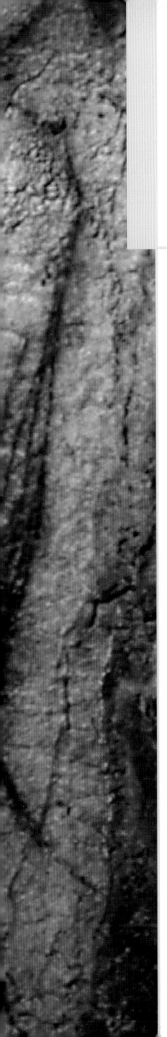

# EXTINCTION

Thirty thousand years ago mammoths thrived across a vast territory encompassing Europe, Asia, and North America and covering millions of square miles. By 5,000 years ago, they had completely disappeared from the planet, except for a small population on a remote Siberian island, which itself died out a thousand years later. What caused the demise of the once widespread mammoths? Various theories have been proposed, based on the apparent influence of humans and on crucial changes in the environment. Only by analyzing and weighing the evidence is it possible to propose a convincing explanation of why, after five million years of evolution, the mammoths finally died out.

*Several magnificent shaggy mammoths form part of the Black Frieze in the cave of Pech-Merle in southern France. Within a few thousand years, this common inhabitant of the Ice Age world had disappeared for ever.*

# CLIMATE AND EXTINCTION

The last woolly mammoths cling to the remaining areas of grassland in this scene set in southern England 14,000 years ago. As the climate became warmer and wetter, the advancing birch forests progressively reduced the extent of the mammoths' habitat and, according to one theory, led to their extinction.

*Modern African elephants* help to maintain their savanna habitat by destroying trees, so preventing the spread of forest. Mammoths, in favorable circumstances, could similarly have helped maintain their steppe and parkland habitats. But they were unable to halt the inexorable advance of forest caused by climatic warming at the end of the last Ice Age.

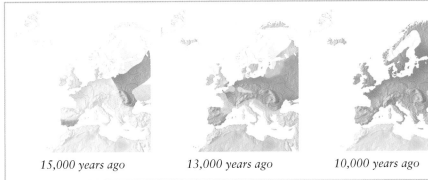

**As the ice retreated** (white), forests (dark green) advanced over Europe from the south and east, replacing the mammoth steppe (pale green). Although 13,000 years ago areas of steppe habitat still remained, they were fragmented and deteriorating, and could have been insufficient to support viable mammoth populations.

15,000 years ago          13,000 years ago          10,000 years ago

***There is some evidence that as feeding conditions changed** for the worse, mammoths became smaller. Examples are seen in the Berelekh remains from eastern Siberia and the Sevsk population from European Russia (see pp.62–3), both less than 17,000 years old and no more than 8 ft (2.4 m) in height. But this effect is not seen everywhere—the later skeletons from Condover, England, and Praz Rodet, Switzerland, are of normal size.*

# HUNTING AND EXTINCTION

A mammoth is caught in a pit trap in southern North America, about 13,000 years ago. Historical elephant hunting in Africa and India has shown that pit traps are an effective means of immobilizing a large animal. However, the archeological evidence for mammoth hunting by prehistoric people is not extensive.

*A rock painting from Ndedema Gorge in South Africa* (right) depicts a Zulu hunter about to hamstring an elephant. His iron axe indicates that the image can be no more than 2,000 years old. Elephants in Africa and Asia were not hunted in large quantities until the expansion of the ivory trade in the 19th century.

**Pitfall traps may have been dug** on paths used by mammoths and camouflaged with branches and brushwood. If the sides were smooth and sloping, converging at the bottom, the animal's legs would be pinned together, leaving it helplessly exposed to the hunters' weapons.

Mammoth extinction involved the demise of two species at roughly the same time—the woolly mammoth *Mammuthus primigenius* across the northern hemisphere, and the Columbian mammoth *M. columbi* in North America. But the extinction of the mammoths was also part of a much wider phenomenon of disappearance of large mammals, which started about 40,000 years ago and reached its peak between 14,000 and 11,500 years ago. A full explanation of extinction must account for all of these losses, not just that of mammoths.

The Late Pleistocene extinction of large mammals was the most recent in a series of periodic extinction events apparent through the fossil record. The most celebrated is the demise of the dinosaurs, sea reptiles, and other creatures around 65 million years ago.

## "Numerous other large mammals became extinct at about the same time as mammoths"

There are many theories for the cause of the dinosaurs' extinction, ranging from an asteroid impact to intense drought. These theories, hotly contested, relate to events in the very distant past. The extinction of the mammoths and other Pleistocene creatures is much closer to us in time, so there is a greater wealth of detailed information available, and a better chance of unraveling the factors that might have been responsible.

To understand the cause of mammoth extinction, it is essential first to establish as accurately as possible when it occurred, and then to look at what was happening in the world at that time. The chief technique for determining the timing of extinction is radiocarbon dating, which gives, within a certain range of error, the date at which a particular specimen was alive. Raw radiocarbon dates, as traditionally quoted, are younger than the absolute date in calendar years. For example, the radiocarbon "date" of the end of the last Ice Age is 10,000 years ago, but the real date is about 11,500 years ago. It is now possible to correct radiocarbon dates by a process known as calibration, and all dates given in this chapter (and throughout the book) are calibrated dates corresponding to the "real" calendar age (see p.170).

Radiocarbon dating of the most recent mammoth fossils from around the world indicates that mammoths did not die out everywhere at exactly the same time. In Eurasia the southernmost populations, such as those in southern Europe and China, may have died out as early as 24,000 years ago. In the rest of Europe, and much of the range eastward across Siberia, the mammoths disappeared at about 13,800 years ago, although in some areas they survived for between one and three thousand years more. Among the latest remains from Europe are skeletons from Condover in England and Praz Rodet in Switzerland, each around 14,200 years old, while the youngest dates from the Berelekh mammoth "cemetery" in eastern Siberia are of similar age, close to the time of disappearance in that area.

The last known continental populations lived in the northernmost parts of Siberia—the Taimyr Peninsula and surrounding areas—where they survived until about 11,150 years ago, and in the Kyttyk pensinsula of Chukotka, in the far north-east, where a date of 9650 years ago, obtained in 2005, marks the latest known occurrence of a mainland mammoth anywhere in the world. Thus, in Eurasia, there seems to have been a general shrinking of the range northward before final extinction, with a particularly marked contraction around 13,800 years ago, although it is possible that some more southerly pockets hung on for a while.

In North America, the latest woolly mammoth dates from Alaska indicate survival until around 13,300 years ago. Further south, data available for Columbian mammoth suggest a somewhat later terminal date of around 12,850 years ago.

Various theories have been advanced to account for the extinction of the mammoths and other Late Pleistocene mammals. An extraterrestrial cause, such as an asteroid impact, is highly unlikely, as decades of intensive geological research into this relatively very recent interval have yielded no compelling evidence for any such event. The only two theories in serious current contention are climatic and vegetational changes on the one hand, and the impact of humans on the other. The timing of mammoth extinction shows that it disappeared across almost all of its range roughly between 14,000 to 10,000 years ago. It happens that two major global revolutions were occurring at that time. First, there were dramatic changes in climate and vegetation as the last Ice Age drew to a close. Second, human populations were rapidly expanding and developing various new technologies, including

# THE GLOBAL EXTINCTION OF LARGE MAMMALS

Between 40,000 and 5,000 years ago many of the world's larger mammals died out. North America lost some 40 species, or 70 percent of the total. In addition to the mammoths, losses included various species of ground sloths, camels, and deer. South America lost 80 percent of its large mammals, including horses, giant armadillos, and giant rodents. In Australia, more than 40 species, or 90 percent of the total, died out, including giant kangaroos, giant wombats, quadrupedal marsupials as big as rhinos, and the marsupial "lion." In Europe and northern Asia there were fewer large mammals to begin with but, in addition to the woolly mammoth, the woolly rhinoceros, cave bear, and giant deer all became extinct.

Almost all of the species that vanished at this time were mammals weighing 90 lb (40 kg) or more; in general, the larger the species the more vulnerable it was to extinction. Thus, in northern Eurasia and North America, all species above a ton in weight were lost. In the middle size range—deer, antelopes, and the like—some species survived while others died out. The smallest species—mice, shrews, and so on—almost all survived.

There are two main reasons why larger mammals are more vulnerable to extinction, by whatever cause. First, they tend to have smaller populations and are therefore more vulnerable to extinction. For example, a given area of habitat might have supported, say, 10 mammoths and 10,000 voles. A reduction in population size of both species by 90 percent would leave only one mammoth but 1,000 voles. The mammoth population would die out, while the voles would survive.

Second, large mammals have a much slower rate of reproduction and so find it more difficult to recover from any decimation of their numbers. A single pair of voles can, in a few years, give rise to several generations comprising hundreds of individuals. Mammoths, on the other hand, were unlikely to reproduce before the age of about 15, had a gestation period of nearly two years, produced only a single calf per litter, and a female would be unlikely to give birth again for three or four years. Their rate of population growth was therefore slow, and a series of local die-offs—due either to hunting or to climatic change—could lead ultimately to extinction.

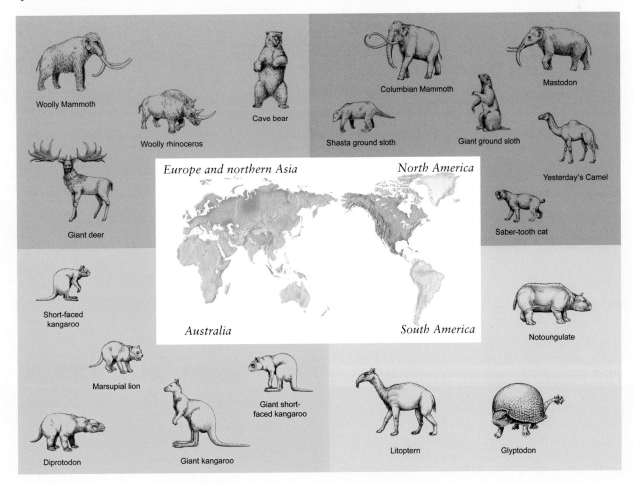

Woolly Mammoth

Woolly rhinoceros

Cave bear

Giant deer

*Europe and northern Asia*

*North America*

Columbian Mammoth

Mastodon

Shasta ground sloth

Giant ground sloth

Yesterday's Camel

Saber-tooth cat

Short-faced kangaroo

Marsupial lion

Diprotodon

Giant kangaroo

Giant short-faced kangaroo

*Australia*

*South America*

Notoungulate

Litoptern

Glyptodon

## THE TRAIL TO EXTINCTION

The woolly mammoth's range contracted in stages from its heyday in the middle of the last Ice Age, until final extinction on remote Arctic islands about 4,000 years ago.

*40–14,000 years ago*

*14–10,000 years ago*

*10–6,000 years ago*

*6–4,000 years ago*

those associated with hunting. The fact that these two potential causes coincided complicates the issue and makes the mystery of extinction more difficult to solve. But, on the other hand, it renders the problem all the more intriguing and may even hold the key to the solution.

## THE HUNTING THEORY

The theory that human hunters were responsible for the demise of mammoths and the rest of the vanished megafauna is termed "overkill." To decide whether overkill could have been responsible for the extinction of the mammoth, two questions have to be addressed. Did people hunt mammoths; and did they do so to a degree that could have caused extinction? Evidence from numerous sites, both in North America and Eurasia, has been used in an attempt to answer these questions.

### North America

Proponents of the overkill theory point to the relatively sudden extinctions in North America that happened roughly 13,300–12,800 years ago, closely following the appearance of human populations who are thought to have entered the continent across Beringia and then swept south. These big-game hunters, with their characteristic hafted, fluted stone spearpoints, take their name from the site of Clovis in New Mexico, where their existence was first recognized in the 1930s.

Certainly the Clovis point was a formidable weapon: modern experiments, using replicas 2–4 in (5–10 cm) long, hafted onto 6½ ft (2 m) wooden shafts with pitch and sinew, showed that when thrown from 65½ ft (20 m) away they penetrated deeply into the back and ribcage of African elephants, and it was found that they could be used up to a dozen times with little or no damage to the point, unless they hit a rib. It is not known whether Clovis people used the spears for throwing or thrusting or both.

There are a dozen sites in the United States where sharpened stone spearpoints of the Clovis type have been found with remains of mammoths, but there is debate among archeologists as to whether these represent hunting of mammoths, scavenging of already-dead carcasses, or accidental association of mammoths and artifacts, by river action for example. One of the most convincing kill sites is at Naco, one of an important group of sites on a stretch of the San Pedro

River in southern Arizona, where eight Clovis points were found with a single adult mammoth: one at the base of the skull, one near a shoulder blade, and five among the ribs and vertebrae.

Other sites show a greater or lesser degree of "association" between mammoth bones and Clovis points, and so provide more circumstantial evidence of mammoth hunting. At Clovis itself, the remains of at least 15 mammoths were found, as well as those of horses, bison, and other animals. The skeletons were

*An excavation at Dent, Colorado, in 1932 (left) produced the first clear evidence for the association of projectile points with mammoth remains in North America. Fifty years later, a large accumulation of mammoth bones was unearthed at Lamb Spring (left, below), also in Colorado, but the evidence of human activity here was more equivocal. The meaning of such assemblages is a subject of debate.*

## "Spearpoints have been found between mammoth bones"

nearly complete, with the bones still in anatomical position, indicating only partial butchering. At the Lange-Ferguson site, South Dakota, one enormous mammoth and a juvenile were perhaps killed and certainly systematically butchered, using tools including some made from an adult mammoth's shoulder blade (see p.132). Three Clovis points were found in an activity area nearby.

Parts of at least seven Columbian mammoths were found at the Colby site, Wyoming, stacked into piles of hundreds of bones; and several stone points, bone tools, and a granite chopper were found with

them. To counter suggestions that the assemblage had been brought together by stream flow, an experiment was performed with African elephant bones, but the investigators could not duplicate the Colby bone piles and argued in favor of a cultural explanation.

Some archeologists have argued that Clovis hunters occasionally killed an entire matriarchal group of females and young. In theory, there are advantages in the culling of whole herds because injured individuals would otherwise be protected by the rest of the herd. If mammoth herds were ever hunted, the most efficient method would no doubt have been a cooperative hunt involving a large group of people. One possible example is at Lehner Ranch, Arizona, where 13 Clovis points were found together with the remains of 13 *M. columbi* calves or young; another is the Dent site, Colorado, which contained the remains of 11 juveniles and young adult females, plus one adult male (see p.149). However, research on tusk growth lines and isotopes from Dent indicates multiple deaths, at separate seasons of the year. If the Dent mammoths were hunted, it was not at the same time, so they could not have been members of the same family group. Many of the Clovis sites, in fact, may represent accumulated deaths over a decade or more.

Studies of modern elephants in Africa highlight the difficulties of interpreting fossil assemblages. Droughts can produce concentrations of elephant bones at water sources; these are dominated by vulnerable youngsters, a fact that could lead archeologists to conclude that a whole family group had been slaughtered there. Humans may still have utilized such carcasses, leaving behind their stone tools, but without hunting the animals or contributing to their demise. Most modern scavenging animals will not or cannot attack a fresh elephant carcass for several days, until it has begun to decompose, so prehistoric people could easily have got there first.

The breakage of mammoth bones may also reflect human involvement, but studies have shown that natural breakage and

*Font de Gaume is one of several caves that contain mammoth figures together with "tectiform" (hut-shaped) forms, interpreted by some as gravity traps or pitfalls.*

*The Clovis point is one of the best-known of prehistoric artifacts. Yet recent dating suggests that the Clovis culture may have lasted no more than 200 years.*

*At the Lynford site in Norfolk, England, a handaxe (right) adjacent to a mammoth toe bone (left) provides compelling evidence of interaction between Neanderthals and mammoths.*

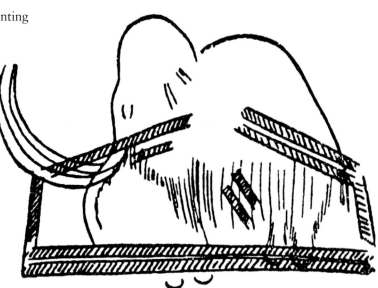

trampling of bones, especially by other elephants, may closely resemble the results of human activities, so caution is required. For example, at the Lamb Spring site in Colorado (see p.149), a bone bed contained numerous mammoth remains (mostly skulls, teeth, and foot bones), as well as those of many other species, together with a battered stone cobble and a quartzite tool. Was human predation responsible, or natural mortality, or both? The different stages of weathering present, the evidence of gnawing on some bones, and the fact that the bone breakage could easily have been caused by the trampling of other animals have led some specialists to see this as a predominantly—or totally—natural accumulation.

## Eurasia

The evidence for widespread mammoth hunting in North America is therefore ambiguous. In Eurasia, there are numerous examples of human utilization of mammoth products, but here again, with a few clear exceptions, direct evidence of hunting is contentious.

One apparently clear indication of prehistoric elephant hunting is the site of Lehringen in Germany, dating to around 125,000 years ago, where a wooden spear more than 7 ft (2 m) long was found in fragments between the ribs of a straight-tusked elephant, *Palaeoloxodon antiquus*, suggesting an early capacity to kill, or at least finish off, a proboscidean.

Other Neanderthal sites provide clear evidence of human utilization of mammoths, but the question of hunting versus scavenging is unresolved. Examples include La Cotte, Jersey (see box below), and Mont Dol, on the coast of western France, where hundreds of juvenile mammoth teeth, dating to around 100,000 years ago, have been found in association with stone tools.

During quarrying in 2002, a channel full of mammoth remains and flint tools was exposed at Lynford in Norfolk, England. The tools (over

# LA COTTE: MAMMOTH HUNTING ON JERSEY

At least one possible episode of mammoth hunting, dating to more than 128,000 years ago, can be seen in a ravine at La Cotte de St. Brelade, on the English Channel island of Jersey. The bones of about 20 mammoths and 5 rhinos were found here in two apparently deliberate heaps, some set up vertically, and some shoulder blades stacked neatly. All the mammoth skulls (shown in dark blue; tusks in white) were smashed open at the back and top, perhaps for access to the brains.

Since the ravine is near the end of a headland, the excavators believe that the animals may have been funneled and stampeded along it and over the edge of the steep fissure, where some fell

more than 100 ft (30m) to their death. The presence of the skulls certainly suggests that, if hunted, they died at this ready-made trap, since there would be no point in transporting the heavy skulls which have little food value. However, it has also been suggested that the mammoths may have fallen in by accident, though subsequently butchered by people.

*A local postage stamp (above) showing the ravine at La Cotte, with prehistoric Jerseymen hauling a mammoth carcass. (Left) A plan of the excavation at La Cotte.*

2700, including 47 handaxes) are of Mousterian type, indicating manufacture by Neanderthals, and this is confirmed by dates of 60–70,000 years ago. It is highly likely that the Neanderthals accumulated the mammoth bones, but hunting is a matter of contention. There are pathologies on the mammoth ribs and vertebrae, including a healed rib fracture that might have been the result of a past hunting episode, although other causes, such as male combat, cannot be excluded. If Neanderthals did hunt mammoths, they probably did so in groups, killing a solitary mammoth with hand-thrusted spears.

Clearly, however, any killing of mammoths by Neanderthals would have been too early to contribute to the extinction of the species. Around 30,000 years ago, Neanderthals were replaced in Europe by modern people, who utilized mammoth products extensively, (see Ch. 4), and persisted through the time of the mammoth's extinction. Their bone accumulations are described by some researchers as "kill sites," although many of them in fact provide evidence of no more than carcass processing. Near Tomsk, in western Siberia, the broken and splintered bones of a young mammoth were found close to a fireplace together with hundreds of stone blades and flakes. At Kraków, Spadzista Street, in Poland, the remains of at least 86 mammoths have been found, dating to about 28-29,000 years ago. The presence of numerous hyoids (tongue bones), all of which bear cutmarks, suggests that the occupants of this site had feasted on roast mammoth tongues. Many of the mammoth bones were arranged into circular piles interpreted as the remains of meat caches. In Germany, at the Gravettian sites of Geissenklösterle and Hohle Fels, the bones of several baby mammoths were found, some with cut marks.

The existence of mammoth-bone huts (see pp.126–9) has led many archeologists to assume that mammoth meat was the staple diet of their occupants. The first major

*A 2-inch (5 cm-) long mammoth bone point manufactured by Neanderthals at Salzgitter-Lebenstedt, Germany.*

bone stockpiles to be found were at Předmostí in Moravia, where the remains of 1,000 mammoths were excavated, with a high proportion of 10–12 year olds. These vulnerable youngsters, just venturing away from the protection of the herd, could have been attractive to human hunters, but equally so to predators such as hyenas, whose kills people might have scavenged.

The presence of gnawing marks on bones in the huts and stockpiles of central and eastern Europe, and of bones of different dates and degrees of weathering within the same structure, suggest that they were collected and scavenged from diverse natural accumulations, and brought in as building materials, raw materials for tools, and as fuel. This, in turn, suggests that mammoths were less significant as a food source than might be thought from the quantity and prominence of their remains in these sites, where reindeer or horse were generally the staples. The occasional mammoth may well have been hunted and killed nonetheless.

Recently, definite evidence of at least one mammoth kill has emerged from Russia. At the swampy site of Lugovskoe, in central Siberia, between the Ob and Irtysh rivers, remains of at least 27 mammoths have been found so far, as well as of 13 other mammalian species, dating to between 19,000 and 14,000 years

*A mammoth vertebra pierced by a spear from Lugovskoe in Russia. In close-up (below), fragments of the stone point remain embedded in the bone.*

ago. In 2002, 300 stone tools were found in deposits at the bottom of the stream, alongside crushed mammoth bones and teeth. The most interesting discovery was a mammoth vertebra pierced by a spearhead made of greenish quartzite, fragments of which remained in the bone. The vertebra is from the mid-chest region, so the strike was probably aimed at the heart, and must have penetrated some 4 in (10cm) of soft tissue and the ½ in- (15 mm-) thick scapula bone. Calculations suggest that the blow was applied with enormous strength, probably by means of a spearthrower used only 16 ft (5 m) away. For the hunter to approach so closely, it is likely that the mammoth was already caught in the sticky mud. The Lugovskoe find represents the first direct evidence of mammoth hunting in Siberia. Some tusks from the site have circular cut marks where they were removed from the skull.

## The Hunting Theory Assessed

While mammoths were clearly hunted on occasion by people, there is no convincing evidence—even in America—for kills on a scale that could have caused the extinction of the species. To save the overkill theory from this problem, the idea of a "Blitzkrieg" has been put forward: following the entry of Clovis people into North America, the mammoths were killed off so quickly that little evidence was preserved. This idea is ingenious but is difficult to prove one way or the other.

## MAMMOTH BUTCHERY

Whether mammoths were killed or scavenged, fresh carcasses were probably not difficult to butcher. In an effort to interpret finds at Olduvai Gorge in Tanzania, where an array of simple tools had been found with early elephant skeletons, it was shown that lava and flint flakes could easily slice through the hide of a dead elephant. In another experiment, a sharpened quartzite flake cut through elephant hide of ¼–⅓ in (5–8mm) thickness. Four sharpenings of the tool were required to cut the whole length of the animal. The same tool easily then stripped the meat from the bones.

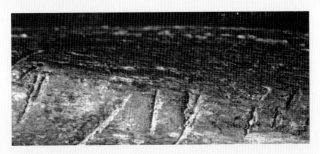

The Efe and Lese people of the Ituri Forest, Democratic Republic of Congo, occasionally kill elephants. Butchery can occupy 25–35 people, who make a short-term camp by the carcass, cut off the meat, dry it in strips on racks, and chop up bones which are boiled for marrow and fat. The carcass is left behind and the meat carried home in baskets. An elephant carcass can be butchered without chopping, cutting or marking the bones. In other words, while the presence of cutmarks on prehistoric bones may well indicate butchery, their absence—as in the majority of fossils—does not rule out the possibility that the carcass was processed by humans.

Four localities in southeast Wisconsin, dated to 16,100–13,050 years ago, show clear evidence of mammoth and/or mastodon butchery. The butchering strategy seems to have been to disarticulate the limbs and feet to obtain sizeable muscle masses, foot pads, and possibly tendons. The limb bones show cut marks, mainly along ridges where muscles attached—interpreted as the result of cutting actions in defleshing. There are also "puncture" marks visible as depressed or crushed bone under the

microscope. These are believed to have resulted from prying or leverage while disarticulating bones.

Few damage marks would be expected on bones from a fresh carcass processed by an experienced butcher. Once the carcass had stiffened, however, more effort would be required to disarticulate and deflesh it, leaving more cut marks. Two of the Wisconsin animals, from Fenske (above) and Mud Lake, showed larger, more frequent cutting and prying marks, and are interpreted as stiffened carcasses that were scavenged. The other two carcasses (Hebior and Schaefer), with many fewer marks, were apparently butchered fresh, whether hunted or scavenged.

*In 1998, researchers reported blood residues on stone tools from northern Alaska, including this point from Girl's Hill. Red blood cells were visible under the microscope, and the team proceeded to analyse proteins and DNA. At four localities, woolly mammoth was the closest match, by comparison with extracts from frozen carcasses. Several of the tools are clearly projectile points—a strong suggestion of mammoth-hunting.*

## POSSIBLE HUNTING METHODS

Clues to the ways in which people may have hunted mammoths may be found in the methods used by native peoples who hunted elephants in recent times. Killing individual animals would presumably have been the most frequent practice, since more would rarely have been required at any one time: a female elephant weighs around 6,610 lb (3,000 kg) and provides 4,000 lb (1,800 kg) of meat.

There are a number of ways to kill a single elephant. The only way to drop one in its tracks is to reach the brain, which would have been difficult for Stone Age hunters. The heart is fairly inaccessible to a thrust or thrown spear, but the lungs make a good target, since even a miss here will hit the intestines. Such wounds may be fatal, especially with poisoned arrows, but only some hours or even days later. A variety of low-risk methods, historically preferred by Africans and Asians for felling these massive, powerful, agile, intelligent, and dangerous animals, includes camouflaged pit-traps; dropping heavily weighted spears on a passing elephant from a tree; and setting foot- or trunk-snares to hold an animal so the hunter can approach and hamstring or disembowel it. Cameroon pygmies cover themselves with elephant dung to mask their scent and crawl from downwind toward their prey until they are close enough to thrust in a poisoned spear or cut a foot tendon.

*Two contrasting pictures depict historical elephant hunting in Africa: an illustration from a book by the British explorer David Livingstone about his travels in Africa in the mid-19th century (above), and a rock art scene from South Africa (below).*

It is also unclear to what extent people relied heavily on mammoth meat during prehistory. Generally they had a broader subsistence base than only hunting the largest herbivores. Both in America and in Eurasia, the peoples of the late Ice Age relied on medium-sized animals—such as deer and bison—as well as on small game and plant foods. New research on carbon and nitrogen isotopes in human bone is aiming to resolve this question (see p.173). The proportions of these isotopes in human bone protein directly reflect those of the diet. Mammoths and woolly rhinos show a carbon and nitrogen isotopic signature different from that of smaller herbivores. The bones of a Neanderthal skeleton from St Césaire, France, suggested that this individual ate a high proportion of mammoth or rhino (whereas hyaenas found at the site were concentrating on horse, bison, and deer). The early modern people from Milovice, Czech Republic, gave a similar result. The investigators cautioned, however, that consumption of freshwater fish could produce a similar signature. More individuals will have to be tested in order to draw any general conclusions about prehistoric diets.

There are also questions over the timing of overkill. Humans had been living in Europe for hundreds of thousands of years without causing megafaunal extinction. In the critical interval toward the end of the last Ice Age, people seem to have returned to central and northern Europe from their Mediterranean refugia around 16,000 years ago. Although more than a millennium before the disappearance of the mammoth from the region, these dates are theoretically consistent with "overkill" during that period. However, there is no clear indication of a technological advance that would suddenly have made the hunting of mammoths more likely or more feasible than previously.

In America, it is becoming increasingly clear that the Clovis people were not the "First Americans;" there is growing evidence, especially from South America, of earlier immigrations, thousands, or even tens of thousands, of years earlier. The cut-marked mammoth bones from Wisconsin (see p.153) pre-date Clovis by up to two millennia. At Lovewell, Kansas, broken mammoth bones, together with a polished bone artifact, have led researchers to suggest human association with mammoth as much as 21,500–22,000 years ago, about 7,000 years before Clovis. However, the overkill theory would still be tenable if the Clovis people in North

*A storage pit at the Russian* bone hut site of Kostenki contains a hoard of mammoth bones. Although some mammoths may have been hunted, the majority of these bones were probably collected from natural accumulations and served as raw materials and fuel.

America, and their contemporaries on other continents, were the first to have hunted mammoths and other big game extensively. Their population sizes were certainly expanding, and the Clovis dates match well. While it has been pointed out that many large mammal species died out before the arrival of Clovis, mammoth was one of the last species to go. The first evidence of people in Alaska is dated to just before 13,800 years ago; mammoth was gone by 13,300 years ago. In the southern United States, the dates are 13,050 and 12,850, respectively.

These dates have been described as a "smoking gun" demonstrating overkill. Opponents point out that many herbivores have been exploited since the Ice Age by much higher human population densities, yet they were not driven to extinction. Native Americans, for example, hunted bison, pronghorns, deer, and wild sheep for millennia without causing their disappearance. In fact, except in cases of human arrivals on islands, it is rare in the archeological record to find people wiping out their prey—when a resource becomes scarce, hunters will generally switch to another and the species will recover.

The inconclusive nature of evidence from "kill sites" has led researchers to look for indirect, and often ingenious, lines of investigation bearing on the question of overkill. Research into the mammoth's growth and life cycle, through the study of tusk rings (see pp.106–07), is one such line of enquiry. For example, a reduction in growth rate of the latest mammoths might be expected if they had died out due to habitat deterioration, but not to hunting. Season of death might also change, from the predominantly late winter to early spring time of hardship, to year-round if death was mainly through hunting. Early trials provide some support to the hunting theory, and although preliminary, they point the way to future research.

In another study, all mammal species of the late Ice Age across the continents were reviewed, and it was found that species that avoided extinction tended to be those that are secretive or nocturnal. As these would have been less visible to human hunters, some support is offered to the overkill theory. The idea that hunters had deliberately targeted the largest animals was rejected by the study, however; all species would have been hunted, but the largest—such as the mammoth—were the most vulnerable to extinction because of their slow rate of recovery.

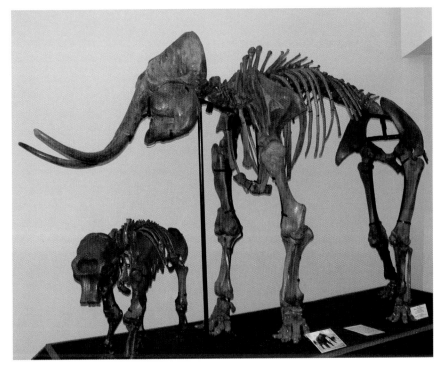

*Adult female and baby mammoth* skeletons from the Lugovskoe site where some, but not all, individuals show evidence of hunting (see pp.152–3). The assemblage spans a considerable period of time, and includes some of the latest mammoths in Asia, south of the Arctic.

The idea that humans precipitated large mammal extinctions, not by hunting, but by introducing disease, has also been proposed. Either people themselves, or domestic animals such as dogs, might have introduced deadly viruses or bacteria. The theory seems unlikely, since no known disease is virulent enough, or broad enough in its scope, to wipe out as wide a range of mammal species as became extinct in, for example, North America. The theory also has difficulty accounting for the extinction only of large mammals. Even if plausible for the Americas and Australia, where people arrived for the first time during the last Ice Age, the theory falls down for Europe and Asia. Here, people had been present for hundreds of thousands of years prior to the extinction of mammoth and other species, so the possibility of a novel disease organism—carried by people—suddenly precipitating a wave of extinction is very low.

Human involvement remains contentious as the cause of the mammoth's extinction. Could climate change have been responsible?

*The treeless tundra of the Aleutian Islands, off the coast of Alaska, shows the kind of landscape that replaced the northern part of the mammoth steppe. According to the climatic theory, this tranformation of the mammoth steppe—which had a richer and more varied vegetation— spelt the end for the mammoth and other large herbivores.*

## CLIMATE AND EXTINCTION

According to the climatic theory of extinction, it was dramatic changes in global climate and vegetation that led to the demise of the mammoths and other large mammals. The maximum extent of the last glaciation, between about 27,000 and 18,000 years ago, saw the expansion of ice caps down to mid-latitudes of Europe and North America. The ice excluded mammoths from a large terrain, and severe conditions reduced their numbers even in unglaciated areas. However, this was not the time when the mammoths and other species became extinct. Areas of the woolly mammoth's steppe habitat remained in the more southern part of the Eurasian range, while suitable vegetation for Columbian mammoths persisted south of the North American ice sheet. The mammoths survived the onslaught of the glaciation, only to succumb to later pressures.

Around 18,000–18,500 years ago global climate began to warm up and the ice sheets started to melt. The warming reached a peak about 15,300 years ago, but this was interrupted by a short period of renewed cold and dry climate between about 12,700–11,500 years ago. The end of this phase marked the true finish of the last Ice Age and the beginning of the modern era, when a prolonged period of relatively mild climate began. Recent research indicates that these changes of climate occurred very rapidly. Around 15,300 years ago, for

EXTINCTION

example, global temperature may have soared by 11°F (6°C) within 10 or 20 years—a far greater jolt than the current increases attributed to greenhouse warming.

This series of dramatic climatic changes occurred at the appropriate time to be a strong candidate for the reduction in range of mammoths between 14,000 and 10,000 years ago. However, supporters of the climatic theory do not generally believe that it was the direct effects of the weather that caused the animals to die out. Woolly mammoths, though adapted to life in a cold climate, could survive through milder episodes, as shown by their persistence through earlier warm phases within and before the last Ice Age. Nor was the Columbian mammoth, with a range extending at least as far south as Mexico, dependent on cold conditions and unable to cope with the heat. Rather, the theory suggests that changes in climate caused major shifts in the pattern of vegetation around the world, and these changes in turn led to the demise of many herbivorous mammals, such as the mammoth, which depended on particular plant foods. Carnivorous mammals also died out because their prey species had disappeared.

*South of the tundra stretches a vast coniferous forest known as taiga, seen here in its fall colors near Vladivostok. Such forests gradually took over the southern part of the mammoth steppe, but were unsuitable habitats for mammoths and other large grazing animals.*

There is much evidence, from fossil plant remains, that vegetational changes at the end of the last Ice Age were profound. Across the northern parts of Eurasia and North America, the vast expanse of grass-dominated, steppe-like vegetation, which had supported the woolly mammoth and many other species, was gradually squeezed out. This occurred because the

## "Changes in climate caused major shifts in the pattern of vegetation"

cool, dry climate that had favored it changed to one of greater warmth and moisture. The increased warmth also melted ice caps, and as sea levels rose, the size of the continents was reduced, raising precipitation in their interiors. The milder, wetter climate encouraged the spread of forests. In the far north, thawing of ice led to waterlogging, which, together with increased snow cover and cloudier skies, reduced plant growth to a tundra condition.

In this way, the mammoth steppe was replaced by landscapes of boggy tundra in the north and coniferous forest in the south, which still exist today. Today's tundra is slow growing and poor in nutrients, and is capable of supporting only limited numbers of specialized mammals feeding on lichen and moss, such as reindeer and musk oxen. The forests, on the

other hand, support tree-browsers such as the moose. Neither of these habitats was suitable for mammoths and other grazing beasts, which were adapted to a diet of grasses and other herbaceous plants.

In North America, south of the ice sheets, the Columbian mammoth also suffered the loss of its natural habitat. The "parkland vegetation" of the Pleistocene, a rich mosaic of grasses, herbs, shrubs, and trees which had provided the Columbian mammoth's mixed diet, largely disappeared. It was replaced in many areas by dense forest; on the open plains by uninterrupted grassland; and in the southwest by semi-desert. Specialist feeders, such as browsing deer in the forests or grazing bison on the prairies, took over.

The reduced diversity of plants in postglacial landscapes may have accounted for the demise of other species as well as the mammoth. Numerous herbivores had existed side by side in the rich mosaic vegetation of the Pleistocene, each species taking its own combination of plant foods. There was a longer period of good feeding each year, since some plants produced growth earlier in the season, some later. There had thus developed a complex web of animals and plants similar to those of the African savanna today. In the forest, tundra, and grassland zones of the postglacial, by

contrast, lower plant diversity and a shorter growing season reduced the diversity and abundance of large mammals that could be supported.

Whatever the precise nature of the changes, it seems clear that the mammoth's habitat was progressively eroded by the new vegetation belts. To investigate whether this led to the final extinction, researchers have dated hundreds of mammoth remains from Europe, Siberia, and North America in an attempt to pin down the pattern of range contraction (see p.148). The loss of mammoths at around 13,800 years ago across much of Europe and Siberia appears very likely to have been triggered by major vegetational change, as it corresponds closely in time to the invasion of forests to the south and the restriction of the mammoth steppe grasslands to the far north. This would explain why the mammoth's range gradually contracted from south to north and the last enclaves finally died out in northernmost Siberia. In some areas mammoth populations may already have been vulnerable to these changes. In Europe, finds of mammoths in the period after 18,500 years ago are rarer and more restricted in range than in earlier times, perhaps because their populations never recovered from the effects of the glacial advance 27,000–18,000 years ago.

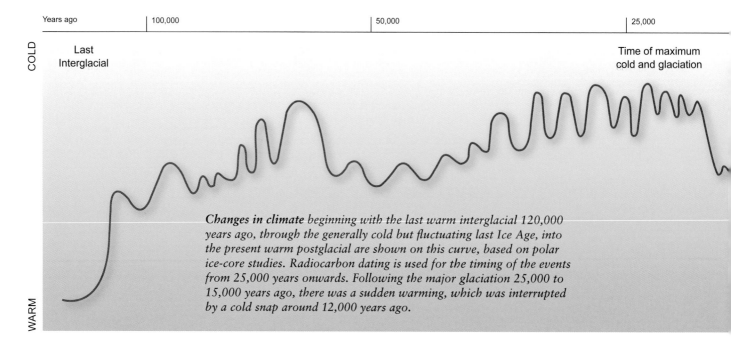

Mammoth range
starts to contract

Years ago    100,000    50,000    25,000

COLD

Last
Interglacial

Time of maximum
cold and glaciation

WARM

*Changes in climate* beginning with the last warm interglacial 120,000 years ago, through the generally cold but fluctuating last Ice Age, into the present warm postglacial are shown on this curve, based on polar ice-core studies. Radiocarbon dating is used for the timing of the events from 25,000 years onwards. Following the major glaciation 25,000 to 15,000 years ago, there was a sudden warming, which was interrupted by a cold snap around 12,000 years ago.

In Alaska, the cold, dry mammoth steppe began to give way at 15,300 years ago, with a transition to a warmer, wetter climate and the expansion of willow, sage, and other woody plants. Grasses and sedges were still abundant, and the mammoth remained. Soon after 13,800 years ago, however, the modern vegetation of tundra and coniferous forest became established—and by 13,300 years ago the mammoth was extinct in the region.

Different mammal species varied in the timing of their range changes, and this provides a further test of the climatic theory. In Europe, mammoth subsisted in the steppic environments between 18,500–13,500 years ago, then promptly disappeared as steppes were invaded by woodland. The giant deer (*Megaloceros*), a species more at home in warmer, lightly wooded environments, increased in abundance just as the mammoth was dying out, enjoying a heyday between about 13,800–12,700 years ago. It then disappeared from Europe with the brief re-establishment of cold, tundra-like conditions. The opposite patterns of these species, corresponding to their differing habitat requirements, strongly suggest that vegetational changes were behind the range contractions of both.

The complex pattern of the mammoth's range reduction over tens of thousands of years shows that the decline to extinction was a gradual process, populations dying out in different areas at different times over a prolonged period. Evidence in support of this idea is coming from studies of DNA preserved in fossil bones (see pp.40–43). Mammoths (as well as other species such as bison and brown bear) possessed considerable

## "Why did mammoths die out only at the end of the last Ice Age?"

genetic variation early in the last Ice Age. As time went on, this seems to have gradually declined, the result of the dying out of individual populations and a reduction in overall numbers.

One problem with the climatic theory of extinction is that the mammoths and other large mammals died out only at the end of the last Ice Age. There have been at least 22 major climatic cycles in the Pleistocene, and thousands of minor ones, but these did not result in such severe levels of extinction. We know that mammoths survived

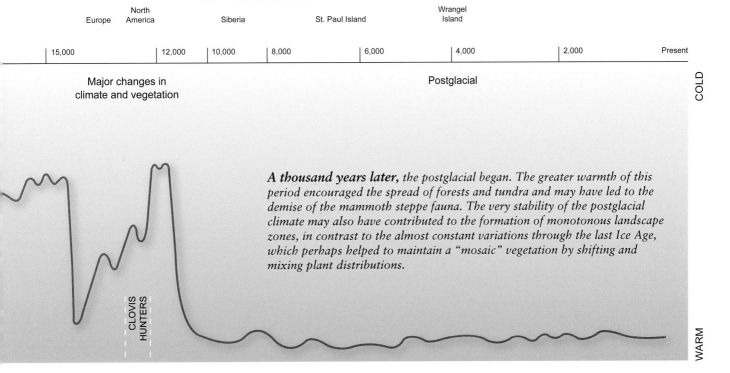

TIMES OF EXTINCTION IN DIFFERENT AREAS

| Europe | North America | | Siberia | | St. Paul Island | | Wrangel Island | | | |

| 15,000 | 12,000 | 10,000 | 8,000 | 6,000 | 4,000 | 2,000 | Present |

COLD

Major changes in climate and vegetation

Postglacial

CLOVIS HUNTERS

*A **thousand years later**, the postglacial began. The greater warmth of this period encouraged the spread of forests and tundra and may have led to the demise of the mammoth steppe fauna. The very stability of the postglacial climate may also have contributed to the formation of monotonous landscape zones, in contrast to the almost constant variations through the last Ice Age, which perhaps helped to maintain a "mosaic" vegetation by shifting and mixing plant distributions.*

WARM

previous interglacial periods of warm climate; why then could they not survive into the present interglacial? This is a powerful objection to the climatic theory, but counterarguments have been raised.

Some claim that although previous interglacials were as warm as today, their vegetation was not so thoroughly zoned into forests and tundra. Instead, at least in some areas, it retained a "mosaic" character, supporting a great variety of plants—the type of habitat that mammoths and other herbivores needed to survive. In the last interglacial, about 120,000 years ago, there is evidence for the persistence of steppic vegetation in at least some areas of Siberia—a refugium for the mammoth that does not exist today.

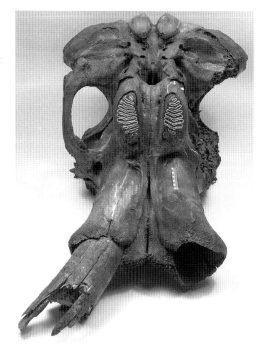

*The skull of the Huntingdon, Utah, mammoth. At roughly 13,000 years old it was one of the last of the Columbian mammoths.*

It may also be that while earlier mammoths could tolerate interglacial conditions, the last populations had become too restricted to cold steppe habitats as a result of evolutionary specialization, and so could not survive into the postglacial. Yet, while this is conceivable for the mammoth, it can hardly account for the extinction of all the species that died out worldwide at the same time.

Another objection to the climatic theory suggests that favorable habitats for mammoths did remain in the postglacial period. So, if that was the case, what prevented them from surviving there? Suitable candidates are the productive grasslands of the American prairies and Russian steppes. On the face of it, these might have supported mammoths. However, some researchers have countered that modern grasslands—unlike the mammoth steppe—are monotonous and dominated by only a few grass species. Mammoths, whose metabolism required a varied diet, with supplements to pure grass, would have found such habitats less favorable and would have lost out in competition with grazing specialists like bison. This argument is less easy to sustain for the Columbian

mammoth, with a southern distribution and a diet including more shrub and tree browse (see pp.89–90). There seem to be at least some places today where this kind of mosaic vegetation has survived naturally—why are the mammoths not still living there?

The discovery that mammoths did survive—albeit in very restricted areas around the Arctic and Bering Seas—into the present interglacial (see p.163), has shifted the debate somewhat, in that the final extinction occurred during a relatively stable climatic interval, thus to an extent weakening the case for a climatic cause. However, if climate were responsible for mammoth loss everywhere else, leaving only tiny, marginal populations at risk of dying off in any event, it would still have to be regarded as the major driver of extinction.

That extensive animal extinctions can occur without human involvement is abundantly evident from the rest of the fossil record; for example, the dinosaurs died out long before people had appeared on the scene. Crucial to the climatic theory of late Pleistocene mammal extinctions is the supposition that the changes at the end of the last Ice Age were different from those of previous glacial cycles. But while the last glacial/interglacial transition is understood in great detail, knowledge of previous ones is much less complete. Researchers will need to demonstrate in greater detail how the last transition was unique, to make a watertight case for the climatic theory of mammoth extinction.

## A COMBINATION OF FACTORS?

Drawing together the threads, it is clear that the problem is not as simple as it might have seemed, and the answers are complex. Two different species of mammoth were involved, extinction happened at different times in different areas, and the environmental and human pressures varied from place to place across the mammoth's vast range. Various factors may,

EXTINCTION

# KILLING MAMMOTHS BY COMPUTER

Mathematical modeling has been used by researchers to examine the possible effects of hunting and climate change on mammoth populations. Computer simulations suggest that a population of mammoths could become extinct in just a few hundred years due either to environmental deterioration or to a relatively low level of hunting. The impact on the mammoths reaches a maximum when environmental and hunting pressures are combined.

The most recent simulations take North America as a test case, and include many mammal species together with the mammoth. Key factors can be varied to assess their impact. These include numbers of humans and their rate of increase, amount of meat eaten per person, numbers of prey and their ranges, reproductive rates of prey, and so on. The researchers found that under a range of scenarios, extinctions result, typically several hundred to a thousand years after human appearance. The closest fit to observed extinctions is obtained at human densities of 3–8 people per 40 square miles (100 square kilometers), with an average of 10 percent of their diet provided by big game. These assumptions, while reasonable, are difficult to verify: the numbers of large mammals at the outset can only be speculated, and some archeologists suggest human population density up to ten times lower even at the peak of Clovis times.

*A replica Clovis point* made from Texas chert and hafted to a cherrywood foreshaft using animal sinew. This would have been inserted into a much longer mainshaft, and either thrusted or thrown with an atlatl (see p.119).

The results, while illustrating the potential impact of human predation, also omit complex behavioral and other factors, on the part of both hunter and hunted, that may have mitigated the outcome. For example, mammoths would move between areas, and groups could be kept going by immigrants from elsewhere. They also, to judge from living elephants, would very rapidly have become wary of human hunters and learnt to avoid them wherever possible. Hunters themselves would have moved to a fresh locality, or switched to other prey, if mammoths had started to become scarce.

Other regions have not been modeled. While human densities in Europe may well have been sufficient to threaten large mammal populations on the above criteria, the vast area of Siberia, from most of which mammoths disappeared by 13,800 years ago, has produced evidence of only sparse human occupation until well after that date.

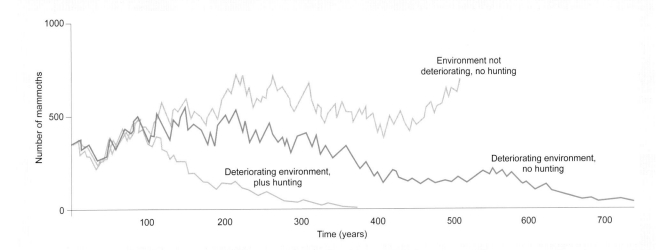

therefore, have combined to produce the total global extinction of the lineage.

Although it appears certain that humans did hunt mammoths, there is no direct evidence that this was more than an occasional activity. The arguments that climatic and vegetational change would have stressed mammoth populations and squeezed them into ever-smaller areas of habitat ("refugia") seem convincing. However, this presumably happened during previous climatic cycles, yet the mammoth pulled through, and the claim that there was nowhere at all that these or any of the other extinct species could have survived in the present interglacial leaves room for doubt whether they died out entirely by "natural causes."

Several researchers have suggested that climatic and human factors may have combined to force the extinction of the mammoths and other large mammals. As their habitat gradually disappeared, patchy mammoth populations clung to remaining areas of mammoth steppe or parkland vegetation. Local factors may have contributed: for example, the western United States may have been affected by drought around 12,900 years ago, perhaps resulting in mass deaths of animals as they crowded around the last waterholes. In these situations, the animals would have been easy prey for prehistoric hunters, perhaps following tracks to locate the isolated

*The exploration of Qagnax Cave on St. Paul Island in 2003 (see box on opposite page). The scatter of bones included the latest mammoths known from North America.*

## "Only concerted and continuing effort will save the elephants from the same fate that befell the mammoths"

mammoth groups. This scenario is not extinction by massive "overkill": it supposes that landscape change killed off most of the mammoths. The proportion killed by humans would have been relatively small, but it may just have tipped the balance between survival and extinction. According to this idea, it was the very coincidence of climate change and human expansion that was fatal for the mammoth and other species. The closer in time that these factors struck (as in North America), the more severe and sudden the wave of extinction. Other researchers conclude that different factors were predominant in different areas—specifically, hunting in North America, but climate and vegetation in Eurasia.

Under any scenario, the key to the mammoths' last days was a progressive shrinking into small,

isolated populations. Many such terminal mammoth groups simply became too small to sustain themselves reproductively and expired; others may have been finished off by humans. Their fate undoubtedly varied from region to region. Island populations fit this pattern: the extinction of the dwarf mammoths of the California Channel Islands (see pp.38–9) closely correlates to the arrival of humans around 12,900 years ago. While the loss of the St. Paul (Pribilof) population seems to have been through habitat loss, the Wrangel Island mammoths may well have expired due to human incursion (see box on opposite page)—and these, as far as we know, were the last mammoths on Earth.

However, the question of whether the last mammoth, or even the last population, died of starvation or hunting, oversimplifies the extinction issue. Extinction was the end-product of a complex process played out over vast terrains and thousands of years. The likely involvement of both vegetational change and hunting issues a stark warning for the mammoth's surviving relatives, the living elephants. They, too, are threatened by the combined pressure of habitat loss (in this case due to human destruction) and poaching; they, too, have seen their once large populations squeezed and divided into thousands of threatened enclaves. Only concerted and continuing effort will save them from the fate that befell the mammoths.

# THE LAST MAMMOTHS ON EARTH

The belief that mammoths had become extinct by 11,000 years ago was shattered in March 1993 when three Russian researchers announced the discovery of woolly mammoth remains between 10,000 and only 4,000 years old. The fossils, from Wrangel Island in the Arctic Ocean 140km north of mainland Chukotka in eastern Siberia (see pp.38–9), have now been the subject of more than 80 radiocarbon dates. These show conclusively that the small mammoths had survived the extinction of mainland populations by thousands of years. While the Egyptian pyramids and Stonehenge were being built, mammoths roamed the Arctic island of Wrangel. It may be significant that the latest mainland mammoths so far dated, at around 8,700 years old, are only 250 miles (400km) from Wrangel, which is believed to have become isolated by sea level rise around 9,500 years ago with its cargo of mammoths on board.

*An evocative remnant of a vanished species, this mammoth tusk was exposed in stream gravels on Wrangel Island, northeast Siberia.*

The reason for their survival may lie in the exceptional vegetation of the island, which even today hosts a much greater variety of plant species than are found on the mainland. Although not identical to the Pleistocene "mammoth steppe," grass and steppic plants are abundant among the tundra of Wrangel.

The cause of the Wrangel mammoths' ultimate demise is uncertain. Recent studies suggest that there was little change in climate or vegetation through the postglacial, including the time of extinction. Clearly, however, they were hanging on under marginal conditions and, given the small size of their population, any changes in habitat could have spelt disaster. On the other hand, there is the possibility that humans may have been involved. The earliest known human occupation on Wrangel dates to around 3,600 years ago. Although this post-dates the last known mammoth by several hundred years, earlier human occupation of other Arctic Ocean islands is known, so it is quite possible that the Wrangel mammoths and people overlapped in time. The early hunters would have found the small mammoths easy prey, and the animals could neither escape nor replenish their numbers from elsewhere. The Wrangel discovery emphasized the extent of our ignorance due to limited sampling over vast areas. A further surprise was in store when, in 1999, Qagnax Cave ("bone cave" in the Aleut language) was discovered by chance on St. Paul Island. St. Paul is one of the Pribilof Islands, in the Bering Sea off the coast of Alaska. Among many bones and teeth of reindeer, polar bear, and other species, scattered on the cave floor, seven were identified as woolly mammoth. Radiocarbon dating showed their ages to be 6400–6640 years old, easily the latest population from the New World. A specimen from another location on the island gave a date of 8,700–9,000 years ago.

As on Wrangel, vegetational changes could explain the persistence of the mammoths on the island but not on the mainland. Since the end of the Ice Age, thanks to its maritime position, St. Paul has been host to lush coastal vegetation of grasses and herbs, while mainland Alaska was converted to shrub tundra and open spruce-poplar forest. However, as the glaciers melted and sea level rose, the Pribilofs at first separated from mainland Alaska around 15,300 years ago, then gradually shrank and split from an initially much larger island, St. Paul reaching its present size of about 35 sq miles (90 sq km) by 5,700 years ago. This area was probably insufficient to maintain a viable mammoth population, which died out. Volcanic activity may have also affected the mammoths by reducing their food supply. However, human involvement can be ruled out, as it is known that people did not arrive on the Pribilofs until the late 18th century.

# Glossary

Words in SMALL CAPITALS indicate cross-references to other definitions.

**Absolute dating** Any method of dating that gives an estimated age in years (compare RELATIVE DATING).

**Accelerator mass spectrometry (AMS)** A form of RADIOCARBON DATING in which the proportion of different ISOTOPES in the sample is determined by direct measurement of the weight of carbon atoms.

**Adaptation** Any aspect of an organism's anatomy, physiology, or behavior that arose in EVOLUTION because it performed a certain function. The principal force bringing about adaptations is believed to be NATURAL SELECTION.

**Algonkians** Native Americans who occupied much of the northeastern part of the country.

**Antediluvian** Literally "before the Flood," a term that was widely used for ancient periods before written history until it was replaced by the term "prehistoric."

**Alluvium** SEDIMENT deposited by rivers at points along the floodplain.

*Amebelodon* A genus of MASTODONT belonging to the family Gomphotheriidae, native to North America in the late Miocene. It had small upper tusks and large, flattened lower tusks probably used to dig up roots.

**Amino-acid racemisation** A technique for determining the relative ages of fossil deposits, it depends on the slow natural equilibration ("racemisation") of the two mirror-image forms of amino-acid molecules (the building-blocks of protein), following the death of an organism. Most commonly undertaken on mollusk shell.

**Anal flap** A flap of skin lying beneath the tail and covering the anus.

**Ancestral mammoth** A term used here to describe *MAMMUTHUS MERIDIONALIS*.

**Ankylosing spondylitis** A painful rheumatic disease involving the fusing together of vertebrae of the spine, often also involving the pelvis.

**Archaeology** The study of the human past through the systematic recovery and analysis of its remains.

**Artifact** A product of human workmanship.

**Artesian spring** Water rising to the surface from an underground reservoir formed from permeable (water-containing) SEDIMENTS enclosed by impermeable strata.

**Barytheres** A group of early PROBOSCIDEA of elephantine size and build, with two pairs of short tusks in upper and lower jaws. Their remains have been found in the Sahara region.

**Base** The part of a DNA NUCLEOTIDE that provides the genetic code. There are four varieties of base, forming a four-letter alphabet whose sequence in a given gene specifies the structure of a protein. The complementary pairing of the bases on the two chains of the DNA molecule provides the basis for DNA replication. Thus, adenine (A) always pairs with thymine (T), and cytosine (C) always pairs with guanine (G).

**Bas-relief** A sculptured figure in art that stands out from its background because material has been removed from around it.

**Behemoth** Name for a primeval creature mentioned in the Old Testament, from the Hebrew *behemoth*; a possible derivation of the word "mammoth."

**Beringia** The area of land comprising northeastern Siberia, Alaska and the Yukon, plus areas of the adjacent continental shelf including the Bering Strait, which today divides the continents. These areas became joined at various times during episodes of low sea level in the PLEISTOCENE.

**Blade** A long, flat, and narrow flake of stone with parallel sides, particularly associated with the Upper PALEOLITHIC period.

**Blitzkrieg** Literally "lightning war," a term borrowed from the German advances of World War II to denote the supposedly rapid killing off of mammoths and other animals.

**Browser** A HERBIVORE that eats predominantly the leaves of trees and shrubs.

**Calibration** In RADIOCARBON DATING, the conversion of raw radiocarbon "dates" into "real" (calendar) years.

**Caries** Tooth decay caused by the dissolution of hard tissue, leading to a cavity.

**Carnivore** An animal that eats predominantly animal matter. Written "carnivore," it can refer to any carnivorous animal. Written "Carnivore," it usually refers to the group of placental mammals that includes the cats, dogs, hyenas, and bears, some of which (e.g. the giant panda) are herbivorous.

**Cave bear** Large, mostly vegetarian, extinct species of European PLEISTOCENE bear, *Ursus spelaeus*. It coexisted with the woolly mammoth and hibernated in caves, where most of its remains have been found.

**Cement** A dental tissue, not as hard as DENTINE or ENAMEL, which forms the outer surface of the root and fixes the tooth in the jaw. In elephants and mammoths, a layer of cement holds the TUSKS and MOLARS in their sockets. It also fills the spaces between the enamel and dentine ridges of the molar crowns.

**Cemetery** A term used informally to refer to large accumulations of bones of living elephants or fossil mammoths. It originated from the erroneous belief that elephants "choose" to die in certain places. The term "elephant (or mammoth) graveyard" is also used.

**Champlevé** Technique whereby material around a figure (e.g. on bone or antler) is scraped away to create a cameo effect.

**Chitin** The tough flexible substance forming the cuticle of insects and other arthropods.

**Chopper** A simple tool with a sharp cutting edge formed by the removal of flakes from one side of a stone or bone.

**Chukchi** The inhabitants of the Chukchi Peninsula of northeastern Siberia, closest to North America.

**Chromosome** The complex structures within the nuclei of living cells that contain the genetic information (DNA), as well as many proteins.

**Cleaver** A tool with flakes removed from both sides, which has a broad cutting edge.

**Cloning** The exact replication of genetic material. It includes the use of genes from one individual organism to produce all or part of a new individual.

**Clovis** The earliest type of fluted point made by Paleo-Indians, characterized by its symmetry, careful flaking, and the removal of a small "flute" (flake) from its face. The Clovis culture lasted only a few hundred years, starting about 13,000 years ago.

**Clubmoss** A group of primitive non-flowering plants (Lycopsida), similar in appearance to mosses, but related to the horsetails.

**Collagen** A fibrous protein extracted from fossil bone for RADIOCARBON DATING and ISOTOPE analysis.

**Columbian mammoth** The common name for *MAMMUTHUS COLUMBI*.

**Computer simulation** An attempt to imitate, by computer, the course of a given process (e.g. mammoth extinction) to assess the likely outcome from various starting conditions.

**Computer tomography (CT)** X-ray analysis whereby an image of an internal slice of an object (a section) can be obtained without cutting the object open. The sections can then be integrated by computer to produce a three-dimensional reconstruction.

**Crown** The part of any tooth above the root, usually covered by enamel. In mammoths, much of the crown was buried within the jaw until the late stages of wear.

**Cut marks** In archaeology, the marks left on bone and other materials by the use of stone or metal tools. Commonly used to identify butchery.

**Cyclops** In Greek mythology, a character with a single large eye in the middle of its forehead.

*Deinotherium* An extinct PROBOSCIDEAN that extended from Africa through southern Europe to India. Although not a true elephant, it was of elephantine form with a trunk and downcurved lower tusks.

**Dentine** A dental tissue, not as hard as ENAMEL, which forms the core of most mammalian teeth. Elephant and mammoth TUSK (IVORY) is largely made up of dentine. In the MOLARS, dentine fills the alternate spaces between the enamel ridges, the others being filled with cement. In life, dentine contains living cells and nerve fibers.

**Dire wolf** A common species of CARNIVORE (*Canis dirus*) in the late PLEISTOCENE of North

America. Equaling the living grey wolf in size, it was of heavier build with a very large head and powerful dentition.

**DNA** Deoxyribonucleic acid, the self-replicating molecule that stores genetic information. An identical full set of an animal's genes is contained within the DNA of each cell, stored within the CHROMOSOMES. A DNA molecule comprises two chains twisted into a helix, each chain formed from NUCLEOTIDE units linked together.

**Dugong** A species of SIRENIA (*Dugong dugon*) up to 12 ft (3.7 m) long, inhabiting coastal shallows in the southwestern Pacific, Indian Ocean and Red Sea.

**Dwarf** A well-proportioned but very small organism beyond the normal range of size variation in the species. The term pygmy is sometimes used in a similar vein.

**Electron spin resonance dating (ESR)** A method of ABSOLUTE DATING that measures the electrons trapped in a crystal lattice as a result of environmental radiation received over time. It has been successfully applied to mammoth molars.

**Electrophoresis** A laboratory process whereby an electric field is applied to a solution of molecules to be analysed. Their rate of movement across the field is proportional to the weight of the molecules, whose relative sizes can therefore be deduced.

**Elephant** A member of the family ELEPHANTIDAE. Technically, it includes the mammoth, although informally it is often restricted to the two living species. The first known usage of the word was by Homer (and Hesiod), who used it to mean "ivory." The word's origin is unknown, although it may be linked to the Hebrew *eleph* (ox).

**Elephantidae** The family within the PROBOSCIDEA that includes the mammoths and living elephants. In contrast to the MASTODONTS they have ridged molars and no tusk enamel.

**Elephas** The genus of elephant to which the living Asian elephant (*E. maximus*) belongs. Various species evolved in Africa and southern Asia, of which only one remains. The name was once used to describe almost all fossil elephants (the mammoth was first named *Elephas primigenius*), although many of these are now recognized as pertaining to different lineages.

**Enamel** The hardest mammalian tissue, forming the outer layer of most teeth and the cutting ridges of mammoth MOLARS.

**Evolution** The process of change by which different forms of life have arisen over long periods of time. The two main processes are, first, the geographical splitting of a lineage to produce two species where one existed before; and second, the change of form within a lineage that results in new structures and ADAPTATIONS. The combination of the two processes explains how a species resembling an ancestral form can continue to live alongside its altered descendant.

**Extinction** The permanent global disappearance of a species of animal or plant. It may

sometimes describe a disappearance from a particular region.

**Fauna** A community of animals comprising all the species occupying a given area; sometimes applied to a fossil assemblage derived from a living fauna.

**Flake** A sliver of stone or bone removed by striking a core with some kind of hammer.

**Flash Flood** Rapid flooding of an area following very heavy rainfall, often when a structure blocking a river is suddenly removed, or when the ground becomes saturated with water that falls so quickly it cannot be absorbed.

**Floodplain** The part of a river valley subject to flooding. As floodwaters recede, their SEDIMENT load is deposited as alluvium.

**Fluorescence** The emission of light from a substance, triggered by the absorption of a photon. Many minerals and substances are naturally fluorescent.

**Foraminifera** Single-celled aquatic animals that secrete a tiny shell that survives in (and contributes to) sea-floor SEDIMENTS. There are many different species, recognizable by their shell form.

**Fossil** The buried remains of an animal or plant usually preserved within SEDIMENT. The term is sometimes restricted to remains that have become mineralized.

**Gastrointestinal tract** The main digestive canal, comprising oesophagus, stomach and intestines.

**Giant deer** Informal name for an extinct group of Eurasian PLEISTOCENE deer that included the so-called Irish Elk *Megaloceros giganteus*. This species coexisted with the woolly mammoth in Europe and bore enormous, flattened, outspread antlers.

**Glaciation** A period of time, within an ICE AGE, when the polar ice caps extended far beyond their present limits.

**Gomphotherium** A common MASTODONT of the Miocene, spreading from Africa through Europe to Pakistan. As large as an Asian elephant, it bore both upper and lower tusks.

**Grazer** A HERBIVORE whose diet consists predominantly of grass. The definition is sometimes extended to feeders on other low-growing non-woody plants.

**Ground-penetrating radar (GPR)** The use of microwave radiation to produce reflected signals from subsurface structures. It can be used to detect buried items from a few centimeters to several meters below ground.

**Ground sloth** An informal term for a number of New World mammalian species belonging to different families within the Order Edentata. At least four species are known from the North American PLEISTOCENE, all extinct by 12,500 years ago. All bore claws and had simple, peglike teeth; they ranged in size up to 18 ft (5.5 m).

**Guard hair** One of the long, coarse hairs forming the outer, protective layer of the fur.

**Herbaceous plant** A plant whose stem is not woody, which usually dies down after the year's growth and may be edible. Includes many grasses and small flowering plants.

**Herbivore** An animal that eats predominantly or exclusively plant matter.

**Holocene** The interglacial that began 11,500 years ago, and in which we are still living.

**Horizon** A layer of SEDIMENT representing a particular interval of past time.

**Hyoid bones** A series of small bones at the base of the tongue that provide attachment for the tongue muscles.

**Hyrax** A group of mammals (Order Hyracoidea) believed to be distantly related to elephants and mammoths, and now restricted to Africa. They range from 1–2 ft (30–60 cm) in length and are herbivorous.

**Ice Age** A period of time within which extensive GLACIATION occurred. Sometimes "the Ice Age" refers to the whole of the PLEISTOCENE, including both glacial and non-glacial episodes. At other times it refers to a particular period within the Pleistocene, as in "the last Ice Age." Even such periods, however, incorporate both times of genuine glaciation and cool but unglaciated intervals. There were other Ice Ages much earlier in the geological record.

**Ice cap** Thick mass of glacial ice and snow that permanently covers an area.

**Ilium** The largest bone of the pelvis, forming the upper and side parts of the pelvic girdle surrounding the birth canal.

**Incisors** The front teeth, used for biting in most mammals. Primitive placental mammals (e.g. shrews) have three pairs, upper and lower; humans have two pairs; and elephants and mammoths one pair, the TUSKS.

**Interglacial** An extended period during the PLEISTOCENE when the climate was as warm, or slightly warmer, than today, and when forests generally extended into northern latitudes. Interglacials do not account for all the time between glaciations, nor are they used for very short warm phases. Interglacials are generally 10,000–20,000 years long.

**In-vitro fertilization (IVF)** The fertilization of an egg by a sperm outside the body, usually in a laboratory dish or tube. The term "in vitro" means "in glass."

**Isotopes** Varieties of a single chemical element that are the same in their chemical properties but that vary in mass due to differing numbers of neutrons in their atoms.

**Ivory** The TUSKS of certain mammals used as a resource, especially those of elephants and mammoths, but also those of walruses and narwhals (tusked whales).

**Lahar** An Indonesian word used by geologists to refer to a mudflow on a volcano. With the appearance of flowing concrete, they are powerful forces capable of rapidly moving large objects for considerable distances.

**Lamut** A major group of people within the TUNGUS complex, who live in Siberia along the Sea of Okhotsk and rely on hunting and reindeer breeding.

**Loxodonta** The genus of elephant to which the living African elephant belongs. The earliest species are found in African deposits 6–7 million years old. The modern species, *L. africana*, is divided into a savanna form (*L. a. africana*, the largest living land mammal) and a smaller forest form (*L. a. cyclotis*). The name derives from the lozenge-shaped enamel ridges on the molar teeth.

**Mammal** A group of vertebrate animals (Class Mammalia) divided into the monotremes (egg-laying mammals), marsupials (pouched mammals) and placentals (all the others, including the PROBOSCIDEA). Characterized by features such as hair, lactation, and the presence of three middle-ear bones.

**Mammoth** An extinct elephant of the genus MAMMUTHUS. The word is often used to refer implicitly to the woolly mammoth M. PRIMIGENIUS.

**Mammoth steppe** A term, coined by U.S. palaeoecologist R.D. Guthrie, to refer to the landscape of northern Eurasia and North America during various phases of the PLEISTOCENE, especially the last ICE AGE. The vegetation comprised a rich mixture of steppe and tundra plants unlike any community today, and supported abundant mammalian FAUNA. It had a greater steppe component in the south and more tundra in the north.

**Mammut** A genus of forest-dwelling MASTODONT, belonging to the family Mammutidae, which included the American mastodon *Mammut americanum*. It was not closely related to the mammoth but, like it, survived until the end of the PLEISTOCENE.

**Mammuthus** The genus within the ELEPHANTIDAE that comprises the mammoths, characterized by twisted tusks and other features. The name, first applied scientifically by the Englishman J. Brookes in 1828, is simply a Latinization of "mammoth." The various species of the genus are listed with the abbreviation "M." for *Mammuthus*.

**M. africanavus** An early species of mammoth found in North Africa, dating from around 3 million years ago. Named by the Frenchman C. Arambourg in 1952.

**M. columbi** The Columbian mammoth, principal mammoth species of southern North America, first named by the Englishman H. Falconer in 1857. It ranged from the northern United States to central Mexico, and survived until around 12,850 years ago.

**M. exilis** The dwarf mammoth of the California Channel Islands. The name, given by Americans C. Stock and E. Furlong in 1928, refers to the supposedly "exiled" status of the island mammoths. However, the Channel Island fossils may represent more than one event of isolation and dwarfing, and hence, technically, more than one species.

**M. imperator** Name given to many mammoth fossils from North America and first applied to a tooth from Nebraska by the American J. Leidy in 1858. Colloquially called "Imperial mammoth," its status is unclear. Some of the material may represent a more primitive stage than later *M. columbi*. Other fossils labelled *M. imperator*, however, are best regarded as *M. columbi* itself.

**M. jeffersonii** Name sometimes given to large mammoth fossils, mainly from southern North America, which may represent an evolutionary stage beyond *M. columbi*. The name, in honor of Thomas Jefferson, was coined by the American H. F. Osborn in 1922.

**M. meridionalis** The dominant mammoth in the early PLEISTOCENE of Europe, its remains date to between about 2.5 million and 750,000 years ago. Called here the ancestral mammoth, it has also been termed "southern elephant" because its remains were first recognized in Italy. The name *meridionalis* is Latin for "southern," and was first coined by the Italian F. Nesti in 1825. Some specialists have placed the ancestral mammoth in a separate genus, *Archidiskodon*, while others divide this into an early species, *A. gromovi*, and a later one, *A. meridionalis*.

**M. primigenius** The woolly mammoth, a species that arose in north-east Eurasia some time between 800,000 and 400,000 years ago and spread widely across the northern hemisphere until its disappearance from the continents about 11,000 years ago. The original name, *Elephas primigenius*, derives from the Latin meaning "first elephant," and was coined by the German J. F. Blumenbach in 1799.

**M. subplanifrons** The earliest species of mammoth, known from fragmentary fossils 4–5 million years old found in southern and eastern Africa. Named by the American H. F. Osborn in 1928.

**M. rumanus** The earliest species of mammoth to appear north of Africa, living in Eurasia between 3.5–2.5 million years ago. It was first named by the Romanian H. Stefănescu in 1924, and later became regarded as merely an early form of *M. meridionalis*, but has recently been reinstated as a result of its distinctly earlier and more primitive nature.

**M. trogontherii** A species of mammoth intermediate in time and anatomy between *M. meridionalis* and *M. primigenius*. Known as the steppe mammoth, it was exclusive to Eurasia and lived between about 1.6 and 0.2 million years ago. Coined by the German H. Pohlig in 1885, the name derives from the fact that the species was first recognized in deposits that also contained the extinct beaver *Trogontherium*. *M. trogontherii* is sometimes given another name, *M. armeniacus*.

**Manatee** Four species of SIRENIA (genus *Trichechus*), 7–14 ft (2.1–4.2 m) long, that inhabit shallow coastal waters, estuaries, and rivers of the tropical Atlantic.

**Marsupial** A group of mammals, now restricted to Australia and the southern Americas, in which the young are born at an early stage of development and complete their growth in the mother's pouch (marsupium).

**Mastodont** An informal term for various lineages of extinct PROBOSCIDEA. The word, meaning "breast-tooth," is derived from the hemispherical cusps of the molars. An alternative spelling, mastodon, is sometimes used, especially for the American mastodon *Mammut americanum*.

**Matriarch** In elephant society, the dominant female of a family group, leading the group in its activities. She is usually an elder relative of many of her subordinates.

**Megafauna** The larger animal species within a FAUNA. In the context of PLEISTOCENE mammals, species over 90 lb (40 kg) in body weight, about the size of a wolf.

**Microwear** The microscopic scratches left on objects, such as teeth or tools, by use.

**Mineralization** The process by which FOSSILS, while buried in the ground, gradually accumulate inorganic minerals that infill and/or replace the original bone, tooth or shell, making them very hard and dense.

**Miocene** The period of time, between 24 and 5 million years ago, that saw the greatest global spread and species diversity of the PROBOSCIDEA.

**Moeritherium** One of the earliest recognized members of the PROBOSCIDEA, known from 40- to 50-million-year-old fossils from North Africa. Like a small hippo in build, it was amphibious and bore incipient tusks.

**Mitochondrion** Microscopic structures within living cells whose main function is the production of energy. They contain a small loop of so-called mitochondrial DNA bearing about 16 genes. A typical cell may contain 2000 mitochondria, each 1–10 micrometers (thousandths of a millimeter) in length.

**Molars** The cheek teeth of mammoths and other mammals. Technically restricted to the back three teeth in each jaw, numbered molars 1 to 3, which are not preceded in life by milk teeth; those in front are the premolars, which are the vertical replacements of juvenile milk teeth. In elephants and mammoths, the six horizontally-replacing cheek teeth comprise the three milk teeth followed by three molars, the true premolars having been lost during evolution. For simplicity, however, all six teeth in mammoths and elephants are commonly described as "molars."

**Mosaic vegetation** A vegetation comprising a wide range of plant species, often an amalgam of different ecological types. Plant diversity is enhanced by small-scale local variations in species composition.

**Mummy** The popular name for a body that has been treated artificially to preserve a life-like appearance, as in ancient Egypt, where bitumen was used in mummification (the word derives from *moumia*, the Persian for pitch). The term is also often extended to human and animal bodies preserved by nature.

**Musk ox** A species of bovid (*Ovibos moschatus*) adapted to the high Arctic and today surviving only in northern Canada and Greenland. Probably related to sheep and goats, it is 4–5 ft (1.2–1.5 m) in height, with large flattened horns and a dense coat.

**Musth** Periodic change in the physiology and behavior of male elephants when they become unusually aggressive for a period of weeks or months.

**Natural selection** The principal process by which EVOLUTION produces ADAPTATIONS. The natural variation between individuals within a species leads to the differential survival and reproductive success of those individuals better able to feed, attract mates, avoid disease and so on. If these features are to some degree inherited, they will become more prevalent among individuals of successive generations.

**Neanderthal** An archaic form of human (*Homo neanderthalensis*)—named after the Neandertal site near Düsseldorf, Germany—which lived from about 150,000 to 33,000 years ago. Neanderthals had a large brain, massive brow ridges, a receding chin and a heavy muscular build. Now generally regarded as not directly ancestral to modern humans.

**Nitrogen-fixing plants** Plants that contain, in their roots, bacteria capable of converting atmospheric nitrogen into ammonia, from which it can be used to form organic molecules. Most are members of the legume group.

**Nucleotide** One of the units of the DNA chain. It is built from three molecules: a sugar, a phosphate and a BASE.

**Ochre** (red or yellow) Soft varieties of iron oxide minerals such as hematite used as pigment for painting and decoration.

**Optically Stimulated Luminescence (OSL)** A method of dating SEDIMENTS that depends on the liberation by light of electrons that have gradually accumulated in the sediment particles since burial.

**Osteoarthritis** A degenerative joint disease caused by wear and tear of bone and cartilage.

**Ostracods** Tiny marine and freshwater crustaceans with a two-valved ovoid or kidney-shaped shell and jointed feeding appendages. Some 200 living species are known, but 2,000 fossil species have been described, many of them only ⅒₅ in (1 mm) or so long.

**Overkill** Denotes the theory that human hunters were responsible for the demise of mammoths and the rest of the vanished MEGAFAUNA.

**Oxbow lake** A curved lake found on the FLOODPLAIN of a river. It is formed when a loop of the river's meander is cut off and the river adopts a shorter, straighter course. The American equivalent is bayou.

**Ozocerite** Term from the Greek, meaning "odoriferous wax," for a yellow to brown native wax (also known as earth wax or mineral wax) made mostly of solid paraffinic hydrocarbons, and associated with sources of petroleum.

**Pachyderm** An informal term for a thick-skinned quadruped, such as an elephant, rhinoceros or hippopotamus, from the Greek *pakhudermos*, meaning "thick skin."

***Palaeoloxodon antiquus*** An extinct species of elephant restricted to Eurasia. Probably derived from a branch of the African *Elephas* stock, it appeared in Europe about 800,000 years ago, and became extinct around 30,000 years ago. This large elephant species, adapted to warm,

forested habitats, occasionally coexisted with the woolly mammoth.

**Paleo-Indians** Hunter-gatherers in the New World, dating from the late PLEISTOCENE to about 7,000 years ago.

**Paleolithic** The "Old Stone Age," the first period of prehistory covering the time from the first appearance of tool-using humans (about 2.5 million years ago) to the end of the last Ice Age around 11,500 years ago. Paleolithic people lived as hunter-gatherers. The period is divided into three major phases. The Lower Paleolithic is the time of the early humans with their simple stone tools. The Middle Paleolithic is more technologically advanced and coincides roughly with the NEANDERTHALS. The Upper Paleolithic is the period of fully modern humans, and is associated with fine stone and bone tools, and the production of imagery and body adornment.

**Paleomagnetism** The determination of the Earth's magnetic field in the past, by the measurement of remnant magnetism in SEDIMENT particles. The known past pattern of variation and reversal of the Earth's magnetic field aids the dating of sediments from their paleomagnetic signal.

**Paleontology** The study of the fossilized remains of living organisms.

**Parallel evolution** A process of EVOLUTION in two separate species or populations which, from the same starting point, produces similar trends or ADAPTATIONS.

**Parietal art** Literally meaning "art on the walls," the term covers prehistoric works of art on any non-movable surface, including blocks, ceilings, and floors.

**Parkland** A landscape superficially resembling a modern ornamental or recreational park, with areas of grass, and scattered groves of trees and shrubs.

**Pelvis** A large bone attached to the rear part of the vertebral column. It supports the internal organs, provides the point of articulation with the hind limbs, and includes the birth canal. Also termed the "pelvic girdle," it is fused from three bones: the ILIUM, ischium, and pubis.

**Periglacial** An environment characterized by severe frost action and dominated by processes and features influenced by cold climates, such as ground ice, the production of large amounts of weathered debris, and strong winds resulting in much deposition of silt and sand. It is often associated with the area fringing modern and PLEISTOCENE glaciers and areas of PERMAFROST.

**Periodontal disease** Disease of the tissues that support a tooth in the jaw, including the cement and jaw bone.

**Permafrost** An area of land in Arctic or Antarctic regions where the ground is permanently frozen. In many areas, the surface layer thaws in summer.

***Phiomia*** One of the paleomastodonts, a group of early PROBOSCIDEA found in North African deposits of 40–30 million years ago. They were the short-trunked precursors of MASTODONTS.

**Pitfall trap** A camouflaged pit, sometimes with a pointed stake at the bottom, dug on paths much used by prey, or near water sources, in the hope that the animal would fall in, so that it could be killed or left to die.

**Pleistocene** The period from approximately 1.7 million to 11,500 years ago, also known as the "ICE AGE." Generally divided into Early Pleistocene from 1.7 million to 780,000 years ago, Middle Pleistocene from 780,000 to 127,000, and Late Pleistocene from 127,000 to 11,500 years ago.

**Point** Category of stone tools, including pointed tools flaked on one or both sides.

**Point Bar** A low ridge of SEDIMENT that forms along the inner bank of a meandering stream.

**Polymerase chain reaction (PCR)** A laboratory process whereby small amounts of DNA are replicated into numerous identical copies suitable for further analysis.

**Portable art** Artworks small or light enough to be carried, such as carved or engraved ivory.

**Postglacial** Another name for the HOLOCENE.

***Primelephas*** An early genus of true elephant (ELEPHANTIDAE) known from African deposits around 5–7 million years old. It is regarded as close to the common ancestor of the modern elephants and the mammoths.

**Proboscidea** The order of mammals that includes the mammoth and the living elephants. Characterized, except in the earliest forms, by the possession of a TRUNK (PROBOSCIS) and TUSKS.

**Proboscis** Technical term for the TRUNK. The word comes from the Greek *proboskis*, literally "means of providing food."

**Projectile point** The tip of a projectile, made of stone, bone, metal or any suitable material.

**Protozoa** Single-celled animals, both free-living and parasites.

**Quartzite** A dense, hard rock that can produce flaked tools; small rounded quartzite cobbles also made ideal hammerstones for stoneworking.

**Radioactivity** The property of unstable ISOTOPES to decay to another type of atom, emitting subatomic particles in the process. The rate of decay is constant for a given isotope, forming a basis for ABSOLUTE DATING.

**Radiocarbon dating** A means of dating organic matter, including bone and wood. Based on counting the regular decay of isotopic carbon ($^{14}C$) to nitrogen, it is accurate back to about 40,000 years ago.

**Red cells** The most common cell type in the blood, technically called erythrocytes, containing the red pigment hemoglobin. They carry oxygen to the muscles and organs, and remove carbon dioxide.

**Refugium** A restricted area of distribution into which a species retreats when conditions become unsuitable elsewhere.

**Relative dating** Any method of determining the relative age of FOSSILS or SEDIMENTS without knowing their absolute age. The information is expressed as "A is younger than (or older than, or the same age as) B."

**Saber-tooth cat** One of a number of extinct species within the cat family (Felidae), which independently evolved very long, flattened, and curved upper canine teeth, probably used for stabbing or slashing prey. The best-known is *Smilodon fatalis* from the Late PLEISTOCENE of North America.

**Sagebrush** Common North American name for species of the genus *Artemisia*, a member of the daisy family (Compositae). Usually of shrubby habit, they are known in Europe as mugwort or wormwood.

**Saltbush** Common North American name for species of the genus *Atriplex*, known in Europe as orache and belonging to the family Chenopodiaceae. They are annuals, commonly found on bare ground by coasts.

**Savanna** A present-day tropical or subtropical vegetation comprising a grassy plain with scattered trees. Typical savanna is seen in parts of Africa that support a diverse mammalian fauna.

**Scapula** The large, flattened bone forming the shoulder blade.

**Scimitar-tooth cat** An extinct genus of large cat (*Homotherium*) with long, powerful forelimbs and enlarged, flattened, serrated upper canines. Belonging to a different group from the true SABER-TOOTHS, the scimitar-tooth cats lived in the PLEISTOCENE of both Europe and North America.

**Sea cow** A large species of SIRENIA (*Hydrodamalis gigas*) up to 25 ft (7.6 m) in length, formerly inhabiting the north Pacific, but hunted to extinction in the 17th century.

**Sebaceous glands** Glands within the skin of some mammals that produce a glossy, water-repellent covering to the fur.

**Sedge** A large group of plants (Cyperaceae), including the rushes and sedges. They are usually plants of damp conditions, but there are also dry-ground species, some of which were common on the mammoth steppe.

**Sediment** In geology, an ancient deposit made up by the accumulation of small particles of clay, silt, sand, gravel or organic matter. FOSSILS are often deposited within it.

**Shaft-wrench** Also known as a perforated baton, this tool comprises a cylinder of bone or antler with a hole through its thickest part. Probably a device for straightening the shafts of spears.

**Shrub** A woody plant usually less tall than a tree, whose stems divide close to the ground.

**Siberia** A vast area of land, extending from the Ural Mountains in the west (the boundary with European Russia) to the Pacific Ocean in the east. It currently comprises a number of republics within the Russian Federation.

**Sinkhole** A craterlike depression due to rock or soil subsidence, which may act as a natural trap for passing animals.

**Sirenia** A group of herbivorous marine mammals, the closest living relatives of elephants and mammoths. Divided into the DUGONG, MANATEE and extinct Steller's SEA COW.

**Solifluction** The process of mass movement of water-laden soil and SEDIMENT as the result of the thawing of frozen ground.

**Spore** The asexual reproductive cell of non-flowering plants such as mosses and ferns.

**Stalagmite** A mineral deposit precipitated in air-filled caves from seeping waters rich in carbonate.

*Stegodon* A genus of MASTODONT belonging to the family Stegodontidae. Within this group were late MIOCENE forms from which the true elephants are believed to have evolved. Ranging from Africa to southern Asia, some species survived into the PLEISTOCENE and produced dwarf forms on Indonesian islands.

*Stegotetrabelodon* The earliest known genus of true elephant from 7 to 6 million-year-old deposits of northern, central, and eastern Africa.

**Steppe** A flat landscape devoid of forest but dominated by grassland. Refers chiefly to areas of southern Russia; similar landscapes are represented by the American prairies.

**Steppe mammoth** The common name for *Mammuthus trogontherii*.

**Steppe-tundra** An alternative term for MAMMOTH STEPPE.

**Straight-tusked elephant** Common name for *Palaeoloxodon antiquus*.

**Subluxation** When one bone moves out of position relative to another, causing the dislocation of a joint

**Supernumerary molar** An extra molar tooth formed at the back of the jaw behind the normal series.

**Taphonomy** Study of the processes whereby animals, plants and ARTIFACTS are incorporated into fossil or archaeological deposits, especially the factors affecting the composition and completeness of excavated remains.

**Tectiform** A class of apparently non-figurative signs found engraved or painted in PALEOLITHIC PARIETAL ART, named for their supposed resemblance to a roofed hut.

**Temporal gland** An organ that, in living elephants, produces a strong-smelling, oily liquid called temporin. It is particularly active during MUSTH.

**Tethys Sea** A large marine basin that formerly extended from what is now the western Mediterranean to Southeast Asia. Many remains of early PROBOSCIDEA have been found in deposits of its former shores.

**Tomography**—see COMPUTER TOMOGRAPHY

**Trunk** Characteristic feature of PROBOSCIDEA, used in breathing, feeding, drinking, grooming, and general manipulation. Formed from a fusion of the nose and upper lip, it carries the two nostrils from the head to its tip.

**Tundra** The dominant landscape of modern Arctic regions, with frozen subsoil, and a vegetation dominated by slow-growing herbaceous plants and shrubs.

**Tundra-steppe** A term for MAMMOTH STEPPE.

**Tungus** A complex of peoples living in the taiga (subarctic forest) of eastern Siberia.

**Tusks** The enlarged incisor teeth of PROBOSCIDEA. Early Proboscidea had tusks in upper or lower jaws, or both. In elephants and mammoths, only the upper tusks are present, and are formed from the second (side) INCISORS, the middle pair having been lost. With the exception of the tips of the milk tusks, ENAMEL is absent, the tusks comprising solid DENTINE.

**"Venus" figurine** The popular but erroneous name for the small female statuettes of the Upper PALEOLITHIC in Eurasia. They span a period from 28,000 to 14,000 years ago, are made of a variety of materials (but often of mammoth ivory), and depict females of a wide variety of ages and physical types.

**White cells** Various types of unpigmented blood cells (leucocytes) that act in the body's defence system against disease organisms.

**Wolverine** A mammalian carnivore (*Gulo gulo*) of the badger family. Up to 3 ft (90 cm) long and powerfully built, it inhabits Arctic regions of Eurasia and North America, hunting small mammals and birds, and scavenging from kills of other carnivores.

**Woolly mammoth** The common name for *Mammuthus primigenius*.

**Woolly rhinoceros** An extinct species of Eurasian PLEISTOCENE rhinoceros, *Coelodonta antiquitatis*, which coexisted with the WOOLLY MAMMOTH. A large rhinoceros with a shoulder hump and woolly coat, it subsisted largely on the grasses of the MAMMOTH STEPPE.

**Yakuts** A major native people of Siberia, who have expanded from their origins on the middle Lena River to become a large autonomous republic, Yakutia.

**Yedoma** Gently sloping, rounded hills of silt in Siberia that may be composed of up to 80 percent ice. Also used to refer to the SEDIMENT of which they are composed.

**Yesterday's camel** Common name for *Camelops hesternus*, the most abundant of the PLEISTOCENE camels of North America. It was extinct by 12,500 years ago.

# Interpreting the evidence

Recent years have witnessed remarkable advances in geological dating, the reconstruction of past climates and habitats, and the interpretation of fossil remains down to microscopic and molecular levels. As a result, paleontology is fast becoming a high-technology science, and much of the information presented in this book is the result of these techniques, which are briefly described in this section.

## WHAT IS A FOSSIL?

Fossils are the preserved remains or imprints of once-living creatures. Although they are by no means rare, they represent only a tiny proportion of the millions of organisms that have lived. After death, most animals disappear completely by decay or by being eaten. Only a very small number end up in river or lake sediment, enclosed within peat, or otherwise buried. The special situations in which some spectacular mammoth remains have been found—frozen in permafrost, pickled in salty tar, or dried in caves—are even rarer, considering the fossil record as a whole. In almost all other situations, the soft tissues quickly decay so that only the hard parts—the bones and teeth of vertebrates, or the shells of invertebrates—remain.

Even bones and teeth incorporated into the deposit can change or indeed disappear after burial, but such changes depend on the surrounding sediment and the timespan. In the case of bones and teeth, acid conditions will dissolve them away until only the hardest parts, or none at all, remain. By contrast, in sediments whose mineral composition is similar to that of bone itself, such as chalk or limestone, the bones may remain little changed for long periods.

Gradually, however, all bones and teeth are infiltrated by minerals dissolved in groundwater. Mineral salts such as calcium and iron will be deposited in tiny pores within the bone, and eventually the bone substance itself, while retaining its original shape, may be replaced atom for atom by substances such as silica. This process, known as mineralization, increases the weight of the fossil and eventually turns it, literally, to stone. In England's Cromer Forest-bed, for example, fossils of the ancestral mammoth, around 1.7 million years old, are frequently heavy and dense, indicating mineralization, while those of the steppe mammoth, less than half as old, are lighter and less altered from their original state. Although it depends greatly on the geological context, as a rule of thumb, bones begin to appear distinctly mineralized after a million years or so. Few mammoth fossils of later species—woolly and Columbian —have suffered significant mineralization, because they are only a few hundred thousand years old or less.

Excavated bones and teeth display a wide range of colors, which reflect the nature of the groundwater but are of little further significance and give no indication of age. For example, bones buried in peat for only a few hundred years may become stained dark brown, while others which have lain in white chalk for a million years can retain their original pale color.

On the matter of terminology, the use of the word "fossil" is a subjective one. Some people restrict it to animal or plant remains that have become mineralized, but since this process is a gradual one, it would be difficult to draw the line between what is a fossil and what is not. Here, all excavated remains of mammoths, including mineralized and unmineralized bones and teeth, and even frozen carcasses, are regarded as fossils, whatever their age. Traces left by once living animals, such as footprints and dung, are also regarded as fossil remains.

## RE-CREATING THE SCENE

### The example of the Berelekh mammoth "cemetery"

When mammoth remains are unearthed, scientists try to gather as much information as possible from the bones themselves, from local geology, and from other remains associated with the mammoths. This information is used to build up a picture of how the deposit originated, and particularly how the mammoth remains came to be there— a study known as taphonomy. The origins of several important mammoth sites have been briefly described in chapter two; here, as an example, a more detailed account is given of one particular site—the Berelekh mammoth cemetery in northeastern Siberia (see p.62).

At the Berelekh site, mammoth bones have been known for centuries; thousands have been exposed by soil movements and river erosion. The bone-bearing horizon has been traced for a distance of 590 ft (180 m) along the bank of the Berelekh River, but how far it extends into the hillslope is anyone's guess. Consequently, it is impossible to know how extensive the site is, but it has been estimated that over the past 100 years up to 50,000 bones from 200 mammoths could have been naturally washed out and redeposited in the riverbed.

The bone deposits, up to 6½ ft (2 m) thick, comprise layers of frozen peat and icy silt. The bones, often entwined with bunches of black hair, form a number of dense concentrations: one, 54 sq ft (5 m²) by 2½ ft (80 cm) deep, contained no less than 954 bones. The bones lack signs of abrasion, indicating that they were not carried downstream in a defleshed state. On the other hand, muscle, cartilage, and ligaments are largely absent, and many bones—up to 42 percent for some types—seem to have been gnawed by predators, which means that they were exposed at some stage. A further indication of exposure is the discovery of many hatched puparia of meat flies in tusk sockets and other parts of the skulls.

There are fish scales and water fleas in the bone layer, which together with the fine silty sediments indicate that this was originally a calm section of a river, perhaps periodically cut off as an oxbow lake. On the other hand, the way in which the peaty and silty sediments are mixed together, and the fact that there are no intact skeletons, or skulls with tusks still in their sockets, show that the remains had subsequently undergone some very active movements and complex disturbances, such as solifluction (soil creep) and landslides. Moreover, because the deposits and bones have been moved around since original deposition, collecting associated animal and plant remains truly contemporary with the bones has to be carried out with care.

Based on tusk shape, about 75 percent of the mammoths recovered are thought to be female, and in terms of age it was young adults that predominated: about 70 percent are between 10 and 30 years old, while some bones of fetuses and young calves were also recovered. One question for investigators is: does Berelekh comprise a few mass deaths, or a more sporadic collection of individuals and small groups that died over many years or even millennia? Like modern elephants, female mammoths with calves and youngsters tended to stay in small groups of 5 to 15, and adult males usually lived alone, although occasional mixed herds of 100 or more individuals were possible. Seen from that perspective, the Berelekh accumulation could have formed from a mixture of individuals and groups, representing accidents over many centuries.

The interpretation of Berelekh is therefore as follows: over thousands of years, a succession of individual mammoths, and perhaps occasional groups, became trapped or drowned, probably by getting stuck in fluid mud or falling through thin ice. Their carcasses were carried off by the river and came to rest in a meander or backwater. When the water level in this silty reservoir dropped periodically, the corpses were exposed, and the stench attracted flies which laid thousands of eggs, hatching larvae that ate the rotting flesh, thus accounting for the discovery of hatched puparia in the skulls.

At the same time the smell attracted scavengers such as wolverines and wolves, which left their gnaw marks on the bones. As the waters returned, fish ate their share of the tissues. Gradually the remaining bones were covered with a bed of silt deposited by the overlying water, and when the basin was fully silted up, peat-forming vegetation grew on its surface. Periodically, however, soil movements added further deposits on top, and mixed up the existing ones. After some millennia, the Berelekh River began to cut through this "cemetery" in its bank, and the bones started to emerge.

# DATING MAMMOTH REMAINS

Calculating the age of ancient finds is essential for making sense of the fossil record, tracing past changes in the environment, and following the evolution and extinction of animals such as the mammoth. All dating techniques are either relative or absolute. If one layer or fossil is higher in the geological sequence than another, then its relative age is later

(that is, it is younger). Relative dating can also be applied if fossils from different localities are associated with assemblages of other fossil animals or plants whose relative ages are known. This allows the fossils in question to be placed in relative order.

In absolute dating, an age in years can be placed on a fossil or deposit. Some techniques allow direct analysis of a mammoth bone or tooth; others date the sediments enclosing it. Sometimes, an age is obtained from deposits above or below a fossil so the latter is seen to be either younger or older than the dated horizon. If the fossil is sandwiched between two dated levels, its age must fall between the two limits.

## Radiocarbon dating

Many absolute dating methods are based on the radioactive decay of one substance into another. The best-known of these is radiocarbon dating. A small amount of the heavy atomic form (isotope) of carbon, $^{14}C$, exists in the atmosphere and is incorporated into plants and animals during their life. It is unstable, however, and gradually decays into nitrogen at a constant, known rate. After approximately 5,730 years, half the $^{14}C$ in a sample will have disappeared, after another 5,730 years, three-quarters, after a further 5,730 years, seven-eighths, and so on. The older the sample, therefore, the less $^{14}C$ that remains, so that by determining the proportion of $^{14}C$ left in the material, its age can be estimated.

There are two methods of radiocarbon dating. In the first, conventional, method the sample, which is generally many grams in weight, is placed in an instrument that counts the number of radioactive emissions (electrons) given out each time an atom of $^{14}C$ decays into one of nitrogen. In the second, newer, method the amount of $^{14}C$ is compared to the commoner, lighter isotope of carbon, $^{12}C$, by direct measurement in a mass spectrometer. This method, known as accelerator mass spectrometry (AMS), has the advantage that it requires only a gram or less of bone or other substance, allowing the dating of precious fossils or artifacts without destroying the whole specimen. Even pigments in cave drawings are now being dated by the AMS method.

Radiocarbon dating depends on the assumption that the production of $^{14}C$ in the atmosphere has remained constant through time. There is evidence, however, that this has varied somewhat in the past, so that radiocarbon dates differ slightly from the true ages of samples, the discrepancy increasing with age. For example, a "radiocarbon age" of 6,000 years implies a true age of nearly 7,000 years. In previous editions of this book, dates for mammoth fossils and climatic events were given in "radiocarbon years" ago. Recent advances in the conversion of radiocarbon dates into real years ("calibration") allow the calculation of an absolute age, albeit within a margin of error, and these are the values that we now cite.

## Other radiometric methods

Because of its relatively fast decay rate, radiocarbon is valid for dating only back to about 40,000 years ago. Beyond that, other radioactive atoms, with slower decay rates, can be measured. These include isotopes of uranium, which decay to lead and cover the past several hundred thousand years; and an isotope of potassium which has an even slower decay rate, to argon, and extends millions of years into the past.

## ESR and OSL dating

A completely different dating method, particularly relevant to the remains of mammoths, is electron spin resonance (ESR), which in principle can cover remains from the whole Pleistocene and beyond (i.e. up to 2 million years ago or more). ESR dating depends on the fact that a buried fossil or other object will receive a steady dose of radiation from cosmic rays and surrounding minerals, especially from uranium that may be absorbed by the specimen. The radiation results in

*A mammoth molar sampled for ESR dating. A rectangular block has been cut from the tooth using a diamond-wire saw. From an inner surface, a small piece of dentine and enamel has been removed with a dental wheel. This provides a sample of clean, uncontaminated material for analysis, and also allows the tooth to be stuck back together with no external damage.*

the liberation of electrons, some of which become trapped in the crystal structure of the fossil. This liberation occurs cumulatively, so the age of a specimen can be estimated by measuring the radiation input from the surrounding deposit, and the trapped electrons in the fossil. The former is measured using a radiation meter at the fossil site, the latter by irradiating the fossil with gamma rays and measuring the resultant ESR signal in the laboratory. Crystalline hard parts of animals are suitable for ESR dating, and mammoth tooth enamel is ideal because it is so thick. Many fossils of mammoths have now been dated by this method.

In a related method, electrons trapped in the crystal lattice of sand or silt grains are released by stimulating with light—Optically Stimulated Luminescence (OSL). This method depends on the useful property that the luminescence "signal" will have been set to zero the last time the sediment grains were exposed to sunlight—generally at the time they were deposited, whether by wind, river, or other means. Once buried, the signal—due to radioactive decay—builds up in a time-dependent fashion, and the luminescence released by applying light in the laboratory provides an estimate of age, generally up to about 200,000 years ago.

## Paleomagnetism

This method of dating depends on the fact that the Earth's magnetic field has varied in intensity through time, and its north-south orientation has "flipped over" on many occasions, most recently at 780,000 years ago. Many geological sediments contain tiny magnetic particles that tend to align with the Earth's magnetic field at the time they are deposited. After carefully collecting a block of sediment and noting its orientation in the field, laboratory measurements can determine the direction and intensity of the remnant magnetism. Because the sequence of magnetic changes through time is well-known from other, independently-dated deposits (especially from the deep sea), the magnetism of a sediment can give a clue to its age, but only in conjunction with other lines of evidence since, for example, "negative" magnetism (indicating a reversal of the Poles) has occurred at many different times in the past.

## Amino-acid racemisation

This method places fossil samples in a relative time scale, and is usually applied to mollusk shells, providing an independent age estimate of other fossils (such as mammal bones) preserved in the same deposit. It depends on the fact that amino acids, small molecules forming the building blocks of proteins in living organisms, occur in two, mirror-image forms, the so-called right-handed (D) and left-handed (L) isomers. It happens that in living tissue, all amino acids are in their L form, but that after death, these gradually alter ("racemize") so that eventually half are in the L form, half in the D form. This process takes hundreds of thousands of years, so by measuring the ratio of D to L forms, an estimate of relative age is obtained. The speed of the process is dependent on temperature and other factors, however, so it can generally be used only to compare samples from a similar region or setting where these factors can be assumed to be fairly constant.

## Limits to dating accuracy

All absolute dating methods are accurate only within certain limits, imposed by two main considerations. First, uncertainties associated with laboratory measurements mean that dates are accurate to within anything from 1 to 10 percent of the "real" value. Second, sample contamination may alter the measured date; for example, a fossil from one horizon which had been affected by soil water running down from a higher horizon may have incorporated some carbon of a later date than itself and will thus give too young a radiocarbon date. For this reason, samples must be collected and chosen with great care, and thoroughly cleaned and tested for contamination before analysis. In radiocarbon dating, for example, the specimen is taken through an elaborate series of chemical and filtration processes, and only carbon from a protein endemic to the fossil bone itself (collagen) is extracted for dating.

## CLUES TO PAST CLIMATE

Mammoths evolved, lived, and finally became extinct against the backdrop of the dramatic climatic changes of the Pleistocene Ice Ages. Knowledge of these changes is therefore essential for understanding the origins of the mammoth's adaptations, migrations, and ultimate demise.

The broad pattern of warm–cold changes in global climate through the Pleistocene (see p.28) has been reconstructed using long cores taken through Arctic or Antarctic ice-sheets, built up by snow and rain over millennia, and through sediment below the ocean floor built up from fine particles settling over millions of years. Successive slices of core provide information on past changes in global climate.

Climatic information is found in ocean cores because of the existence of two types or isotopes of oxygen in sea water ($H_2O$), one of which ($^{18}O$) is slightly heavier than the other ($^{16}O$). During times of cold climate, proportionately more water containing the lighter isotope evaporated from the ocean surface, and some of it became locked up as ice when it fell as rain or snow on the expanded glaciers. The water remaining in the sea therefore accumulated proportionately more of the heavier isotope. During times of warmer climate, this effect was less pronounced, and the ocean retained more of the lighter isotope. The relative "heaviness" of oxygen from the ancient ocean therefore provides a measure of past temperature.

The oxygen signal has been preserved in the shells of microscopic sea creatures called foraminifera, or forams. In their growth they take in sea water and incorporate the oxygen atoms into their carbonate shells. Thousands of shells of dead forams are preserved in each slice of the sea-bottom sediment, so, by analyzing the proportion of $^{18}O$ to $^{16}O$ within them, scientists can estimate the global temperature at the time that slice of sediment was being deposited.

Cores are drilled and lifted by specially equipped oceangoing vessels. In the laboratory, the cores are divided into slices representing successive intervals of past time. Under the microscope, foram shells are picked from the sediment, and then dissolved in acid to produce carbon dioxide ($CO_2$) from calcium carbonate in the shell. This carbon dioxide is collected and analyzed in a mass spectrometer, an instrument that measures how much $CO_2$ is made from $^{18}O$, how much from $^{16}O$. The ratio between these figures gives an estimate of past temperature and, by plotting these estimates from slices along the whole core, a graph of past climate change is produced.

Similar analysis performed on cores taken through ice sheets (glaciers) in Greenland and Antarctica has provided an even more detailed record of climate, especially over shorter periods (e.g. the last Ice Age—see pp.158–59). As in the ocean cores, it is the proportion of oxygen isotopes that bears the climatic signal. Here, however, it is the lighter isotope ($^{16}O$) that indicates colder climate, because the ice is formed from snow and rain derived from water evaporating from the ocean, comprising relatively more of the lighter form as temperature decreases.

## RECONSTRUCTING THE MAMMOTH'S HABITAT AND DIET

While studies of ocean sediments and ice cores provide invaluable information on broad changes in global climate, other methods are used to give more details of the climate and landscape of particular regions at specific times, including the local habitats of individual mammoth finds. These include the study of other animal and plant remains found in dated horizons, or directly associated with the mammoths themselves.

### Plant remains

Direct evidence of past vegetation is obtained by studying plant remains in ancient deposits. Some such deposits, for instance those formed of peat, are composed almost entirely of plant matter, as are the remains of mammoth gut contents and dung balls. Others, such as silts, or clays from rivers or lakes, must be washed and sieved to reveal and concentrate fragmentary plant remains. Parts that can be recognized include fruits, seeds, and leaves. Under the microscope these can often be identified to the level of family, genus, or species, with the aid of complete specimens of modern plants for comparison. Trees can often be identified by their wood, because different species have a characteristic appearance when thin sections are examined under the microscope.

### Pollen analysis

A particularly important category of plant remains comprises the individual pollen grains shed by long-dead flowers, and the spores produced by ancient ferns and mosses. Although generally only one-fiftieth of a millimeter or so in diameter, these have a tough outer wall which is extremely resistant to decay and erosion, and often survives in ancient sediments. They also appear to survive attack by mammalian digestive enzymes, and so are preserved in gut contents and dung. Moreover, under the microscope, the spores and pollen of different species of plant have clearly differing characteristic shapes, which allows them to be identified.

Samples of peat, silt, or clay from fossil deposits may contain millions of spores or pollen grains which blew or were washed in while the sediment was accumulating. For study, however, they must be separated from the sediment and concentrated, involving a series of laboratory procedures, including flotation, sieving, and dissolving away other plant or mineral matter with various chemicals. Eventually, the pollen and spore concentrate is smeared on a microscope slide and systematically scanned, each identifiable grain being noted. From this, a list of plant species and their proportions is obtained.

Interpreting pollen and spore data requires care. Some plants (e.g. pine and birch) are known to produce much more pollen per flower than others (e.g. beech and holly), so their preponderance in fossil sediments could be artificially inflated. Also, some pollen types, such as pine, can blow farther in

*The varying shapes of pollen and spores, magnified 1,000 times. Clockwise from top left: clubmoss* (Selaginella), *grass* (Graminae), *pine* (Pinus), *and sagebrush/mugwort* (Artemisia).

the wind than others, so their presence does not necessarily mean that there were pine trees in the immediate area. Finally, certain grains (e.g. some moss and fern spores) are more resistant to decay than others (e.g. pollen from poplar trees), so are more likely to survive in ancient sediments.

As long as such biases are borne in mind, however, pollen and spore analysis is a powerful means of reconstructing past vegetation. Combining the data from pollen and other vegetational remains can help overcome any possible bias. For example, the guts of the Shandrin mammoth (see pp.56 and 88) contained abundant moss spores, suggesting that moss was a major part of its last meal. However, the actual plant fragments contained only 1 percent moss by volume, giving a rather more accurate indication of its importance.

## Insect remains

Other organic remains are also used to help reconstruct past climate and habitat. These include mollusks, diatoms (microscopic single-celled plants that live in salt and fresh water), and ostracods (tiny freshwater crustaceans). Insects, especially beetles, have been particularly employed in this way, since their hard outer skeleton, made of the tough substance chitin, survives well in ancient sediments. Usually they are not found complete, but isolated parts, such as the head, wing covers, or genitalia, have detailed structures which, under the microscope, allow them to be identified to particular species, by comparison with known modern specimens. As with plant remains, the insect parts are recovered from sediments by a combination of sieving, floating, and chemical techniques.

Insect remains are particularly valuable for habitat reconstruction because many species survive only within relatively narrow tolerances of climate and vegetation. This is deduced from knowledge of living representatives. For example, some species are restricted to life on a particular plant species, and so their discovery in an ancient sediment indicates the presence of the plant even when remains of the vegetation are absent. Further, many insects have limited temperature tolerances and so their fossils can suggest past climate. Many deposits of the last Ice Age in Britain, for example, contain beetle species that today live in northern Scandinavia or Siberia. Conversely, interglacial deposits include beetles now living only in southern Europe. Detailed analysis of beetle species, in comparison with their modern ranges, can allow scientists to estimate past winter and summer temperatures when the deposit was forming.

## Dietary evidence from teeth and bones
### Microwear analysis

The diet of different mammoth species is broadly reflected in the structure of their teeth, especially the height of the tooth crown and number of enamel ridges (see pp.26–27), but techniques exist for a more direct indication of the food eaten by a particular animal. Study of a range of living mammals of differing, known diets, has shown that microscopic marks left on the surface of tooth enamel are related to diet. Among herbivorous animals, those that graze on tough grasses tend to have the enamel surface scored by long, thin scratches, whereas those that chew on the leaves, bark, and wood of trees and shrubs tend to show shorter, sometimes deeper pits. These marks are visualized under a scanning electron microscope (SEM), but mammoth teeth are too large to put in the machine, so an accurate latex peel of part of the enamel surface is made, coated with a very thin layer of gold, and observed in the SEM. Work on living mammals shows that the marks renew every few weeks in life as the teeth wear down, so will illustrate only the last weeks of diet of a fossilized animal. Results from many individuals are therefore required for gaining a broad picture of the diet of a particular species.

### Isotope analysis

Carbon (C) and Nitrogen (N) isotopes preserved in the molecules of fossil teeth and bones can also provide clues to diet. The proportions of the rare isotope $^{15}N$, compared to the common $^{14}N$, and $^{13}C$ compared to $^{12}C$, reflect the composition of food eaten over the last few years of the animal's life, and can be measured by analyzing fossil samples in a mass spectrometer. In the tropics, grasses have different biochemical pathways from other plants, resulting in an enrichment of $^{13}C$ that can be detected in the remains of mammals that predominantly ate grass. In temperate to Arctic regions, where mammoths mostly lived, the signal is more subtle. Forested conditions are believed to result in elevation of $^{13}C$ levels. $^{15}N$ is enriched in species that eat predominantly animal food, and varies among herbivore species for a variety of reasons, such as the proportion of nitrogen-fixing plants in the diet, as these have a reduced proportion of $^{15}N$.

## PUTTING FLESH ON THE BONES
### Reconstructing body form

Mammoths are almost unique among extinct prehistoric animals in the detailed information available on their soft parts such as skin and hair, thanks to the preservation of frozen carcasses. Nonetheless, the frozen material is shriveled, and no carcass except the baby Dima has reached the laboratory wholly intact. Further evidence on the animal's body form is provided by Paleolithic art and the study of skulls and skeletons.

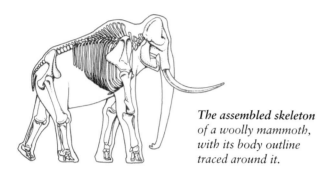

*The assembled skeleton* of a woolly mammoth, with its body outline traced around it.

Piecing together a mammoth skeleton for museum display is a particularly revealing exercise, for then characteristics such as the size of the animal's rib cage and the relative height of its front and hind quarters become evident. Assembling the spine, it becomes clear that the individual vertebrae fit together in such a way as to produce the sloping back seen in Paleolithic art, with the shoulder hump resulting from long vertebral spines.

The shape of the mammoth's head can be deduced from that of the skull, allowing for flesh using the modern elephant as a guide, in essentially the same way as police reconstructions of faces are made from human skulls. The mammoth's skull shows its eye socket to be relatively far forward, for example. Details such as the shape of the ear are based on frozen material. The top of the skull is rather pointed, but it is clear from Paleolithic art that in life this was padded into a rounded dome, presumably with fatty or other tissue.

*The skull of a woolly mammoth,* with its head outline traced around it and other details added.

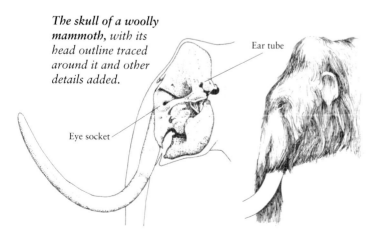

Ear tube

Eye socket

### How to weigh a mammoth

The approximate body weight of a mammoth species can be calculated from its height, by comparison with modern elephants whose body form is broadly comparable. Measurement of elephants has resulted in graphs showing the increase in body weight with shoulder height. For mammoth specimens within the size range of living elephants, body weights can be read directly from these graphs. For example, a mammoth 11 ft (3.4 m) high can be assumed to weigh about 6 tons—about the same as a large bull African elephant of the same height. For smaller individuals, such as dwarf mammoths, juvenile elephants can be used as a guide, with something added for the heavier build and larger tusks of dwarfed adults. Thus, a 6 ft (1.8 m) high adult dwarf mammoth probably weighed around 1.5 tons, slightly above the 1.3 tons of a 10-year-old elephant of comparable height. Conversely, some mammoths, such as *Mammuthus trogontherii* and *M. columbi*, were larger than any living elephant, shoulder heights reaching more than 13 ft (4 m) in large males. In this case, the graph for living elephants can be extended upward, indicating a weight of around 10 tons for these individuals. A check can be made by comparing the girth of the fossil limb bones with those of living elephants, for it is found that these correlate closely to body weight among mammals. By either method, weight is found to increase greatly with relatively small height increments, since a larger animal puts on weight in all dimensions. Weight, in fact, is roughly proportional to the cube of height or length, so that, for example, a doubling in height produces an eightfold increase in weight.

## MAMMOTH AGE AND SEX
### Age estimation

The age of a juvenile can be roughly estimated from its size, but this is not a very accurate method, and cannot be applied to adults that have stopped growing. Much more precise age estimates can be obtained from the teeth. As explained on pp.92–93, mammoths went through six sets of molars in their life, each successive tooth larger than the previous one. By studying large numbers of teeth, researchers learn to place any fossil molar in its correct position in the sequence, based on its size, crown height, and number of enamel ridges. Immediately, therefore, it is possible to get a rough idea of a mammoth's age.

More accurate estimates are made by comparison with modern elephants. Research on African and Asian elephants has shown how the stage of tooth replacement and wear is linked to age. For example, the third molar always replaces the second at about 3–4 years of age, while the sixth replaces the fifth at about 30. These figures can be taken as roughly valid for woolly mammoths too, because the elephants have the same number of tooth replacements as the mammoths, and are of comparable body size indicating a similar total lifespan. Larger species, such as the Columbian mammoth, still had six sets of teeth, but because of their larger size

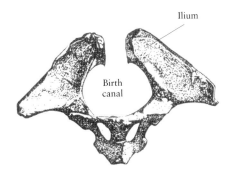

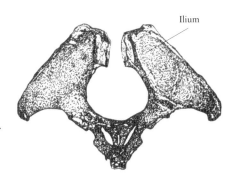

*Pelvic girdles extracted from male and female woolly mammoths whose gender is known from preserved genitalia. In the female (left), the birth canal is wider, and the surrounding bone narrower, than in the male (right).*

probably lived longer. The age at which each tooth was in use may, therefore, have been spread over more years than in the elephants and woolly mammoths, so a small addition to "elephant tooth ages" must be made.

### Male or female?

A few frozen carcasses preserve intact genitalia, so gender can be directly observed. Most usually, however, gender has to be deduced from skeletal remains. Males were generally larger than females, so the sex of very large or very small individuals (provided they were adult) can usually be guessed. However, in many cases of intermediate size it would be difficult to know from the bones whether they belonged to a large female or a small male. Male skulls are relatively more robust than those of females, especially in the region of the tusk. If tusks themselves are preserved, the much more slender tusks of females compared to the stouter male variety (see p.94) may provide a clear clue to gender.

The most reliable method of determining sex is to examine the pelvic girdle. Because this bone contains the birth canal through which baby mammoths were born, its shape differs clearly between males and females. In the female, the birth canal is relatively wider, and the bone surrounding it, the ilium, is proportionately narrow. In males, conversely, the equivalent hole is narrower and the ilium wider. Measurement of a series of skeletons has shown that the ratio of canal width to ilium width is always higher in females than in males.

## GENES FROM FOSSILS: MAMMOTH DNA

Genetic material, or DNA, is preserved in ancient tissues only in special circumstances where decay is minimized, as in very dry conditions or in quickly frozen carcasses. Surprisingly, bones, teeth, and hair may preserve DNA better than soft tissues. The DNA originates in living cells that produce their surrounding hard tissue, and the mineral nature of their environment protects the DNA from decay.

Samples are chosen that show the least contamination (for example, from soil chemicals or handling during excavation), and their surfaces are cleaned and washed. The sample is finely ground, and the resulting suspension is treated with enzymes and solvents. Even under ideal conditions, ancient tissues preserve only minute amounts of DNA, so the DNA must be multiplied into sufficient quantities for analysis.

This process, known as amplification, is achieved by the inherent potential of DNA to replicate itself, using the polymerase chain reaction (PCR). A DNA molecule consists of two long strands, each formed from a chain of smaller molecules called nucleotides. The nucleotides contain a part known as the base, of which there are four varieties, abbreviated as A, T, C, and G. These bases form a sort of genetic alphabet whose order along the DNA molecule specifies the mammoth's genetic code. Moreover, the two DNA strands are linked in such a way that C on one strand always pairs with G on the other, while A always pairs with T. In life, this provides the key to DNA replication, because if the two strands are separated, each provides a template for the growth of a new partner strand.

In the laboratory the double-stranded DNA is separated into single strands by heating. Two "primers" (short pieces of synthetic DNA) are added to this DNA sample, and link up with that part of the animal's DNA which it is intended to replicate. Also present are quantities of the four single nucleotides, which link up to the primer DNA one by one, using the sequence of the mammoth DNA as a template. This process repeats 20–40 times, doubling the amount of DNA with each cycle, producing millions of copies of the original mammoth DNA.

The next stage is to work out the sequence of nucleotides in the mammoth DNA. This is achieved by a complex and ingenious process which can only be outlined here. Quantities of the DNA bases are "labeled" (that is, they have attached to them) a fluorescent dye molecule—a different type for each of the four bases. During replication of the DNA molecule, the labeled bases are added to the solution, and are so designed that they stop the lengthening of the DNA chain whenever one happens to be incorporated. The result is a series of incomplete chains, each of them one base longer than the next. The final products are drawn through a gel-filled capillary by an electric field (the process of electrophoresis). DNA fragments in the sample move along the gel, the time taken by each fragment being proportional to its length. As the identity of the base at the end of each chain can be observed by the color of its fluorescent label, the complete base sequence of the original chain can be determined as the sequence of labeled end-bases in successively longer chains.

# SITES IN EUROPE

NORTHERN IRELAND

Aghnadarragh

Belfast

NORTH SEA

SWEDEN

DENMARK

Hedehusene

Lockarp

ENGLAND

Condover

West Runton

WALES

Lynford

Paviland

Cromer Forest Bed

Stanton Harcourt

Kent's Cavern

ENGLISH CHANNEL

La Cotte de St. Brelade

Mont-Dol

ATLANTIC OCEAN

Ilford and Aveley

Deep Water Channel

Brown Bank

Red Crag

Amsterdam

NETHERLANDS

Eurogeul

Bergharen

Ahlen

Eastern Scheldt

Aa

Zemst

Tegelen

BELGIUM

Gönnersdorf

Polch

Mosbach

Otterstadt

Paris

Mayenne-Sciences

Arcy-sur-Cure

Steinheim

Reseau Guy Martin

La Marche

FRANCE

Las Caldas

El Pindal

El Castillo

El Arco B

Brassempouy

Lespugue

Lourdes

Gargas

Les Trois Frères

Canecaude

Bize

Madrid

SPAIN

El Padul

MEDITERRANEAN SEA

Salzgitter-Lebenstedt

GERMANY

POLAND

Warsaw

Edersleben

Voigtstedt

Süssenborn

Kniegrotte

Prague

CZECH REP.

Obłaz

Brno

Předmostí

Milovice, Dolní

Věstonice, Pa

SLOVAKI

Mátra

Buda

HUNGAR

Obere Klause

Hohlenstein-Stadel

Vogelherd

Geissenklösterle

Hohle Fels

Krems-Wachtberg

Siegsdorf

AUSTRIA

Tata

Praz Rodet

SWITZ

Chauvet

Senèze

Chilhac

Oulen

Le Figuier

Chabot

Durfort

La Baume Latrone

Upper Valdarno

Montopoli

Pietrafitta

Scoppito

Rome

ITALY

SARDINIA

San Giovanni di Sinis

The maps, on this and the following pages, show the position of important mammoth sites described in the text. They represent only a small fraction of the number of sites known. For example, over 250 mammoth sites are known in Texas alone, while only four are shown on the map on p.180.

MAMMOTHS
- Ancestral
- Steppe
- Woolly—skeleton
- Woolly—pickled
- Dwarf

ARCHEOLOGICAL SITES
- Cave depictions
- Portable depictions
- Ivory and bone craft objects
- Mammoth-bone huts

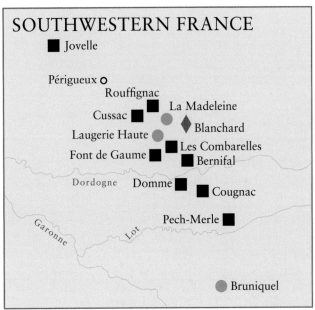

## SOUTHWESTERN FRANCE

Jovelle

Périgueux ○

Rouffignac

Cussac — La Madeleine

Laugerie Haute — Blanchard

Font de Gaume — Les Combarelles

Bernifal

Dordogne — Domme — Cougnac

Garonne — Lot

Pech-Merle

Bruniquel

## ISRAEL & ETHIOPIA

Ubeidiya
Jerusalem

EGYPT

SAUDI ARABIA

SUDAN

RED SEA

Afar Depression

Addis Ababa

ETHIOPIA

# SITES IN EASTERN EUROPE AND ASIA

BARENTS SEA

KARA SEA

FINLAND

Helsinki

St. Petersburg

Yamal Peninsula
(*Mascha*)

Warsaw

Yuribei
Mongochen

Mezin
Yudinovo

Eliseevich

Sevsk

Moscow

Zaraisk

Sungir

Gagarino

Mezhirich

UKRAINE

Avdeevo

?Ignatiev

Lugovskoye

Kapova

Chembakchino

Kostenki

R          U          S

Nogaisk

Kiik Koba

Khapry

Teguldet

Azov

Tomsk

Siniaya Balka

Omsk

Shestkovo

BLACK
SEA

GEORGIA

CASPIAN
SEA

Dmanisi

## MAMMOTHS

- Ancestral
- Steppe
- Woolly—skeleton
- Woolly—frozen
- Dwarf

## ARCHEOLOGICAL SITES

- Cave paintings
- Portable depiction
- Ivory and bone craft objects
- Mammoth-bone huts
- Open-air depictions

Wrangel Island 🐘

Kyttyk Peninsula 🐘

BERING SEA

Olyer Suite 🐘

Liakhov Islands 🐘    Shandrin 🐘
LAPTEV SEA         Sanga-Yurakh 🐘
Taimyr 🐘              Berelekh 🐘    Beresovka 🐘

Jarkov 🐘  🐘
Fishhook    🐘 Khatanga    Lena River
(Adams)     Maxunuokha
            River (Yukagir)

Y A K U T I A    Kirigilyak (Dima) 🐘

Oimyakon 🐘

S I B E R I A

Vilui 🐘

S E A   O F   O K H O T S K

S    I    A

Mal'ta 🔴
Irkutsk                          Nemuro Channel 🐘

Zhalainuouer 🐘

Harbin ☐

Tsagaan Salaa/Baga Oigor ∎    Heilongjiang 🐘

M O N G O L I A                    J A P A N

Maguanjou 🐘        Bejiing ☐

C H I N A                         Osaka Group 🐘

Yushe Basin 🐘        Ji'nan 🐘

# SITES IN NORTH AMERICA

BEAUFORT SEA

CANADA

Elephant Point,
Eschscholtz Bay

Girl's Hill

St. Paul

A L A S K A

Old Crow

Fairbanks Creek

Anchorage

YUKON

GULF OF ALASKA

Wally's Beach    Wellsch Valley

Lange-Ferguson

Colby    Hot Springs

Fenske, Schaefer,
Heboir and Mud Lake

U N I T E D    S T A T E S

Hay Springs

Jonesboro

Huntingdon    Dent
Moab

Franklin
County    Angus
Lovewell

Big Bone Lick

Bechan Cave, Cowboy Cave

Lamb Spring

Rancho La Brea

Miami

Domebo

Santa Rosa
Island    Anza-Borrego

Blackwater
Draw, Clovis

PACIFIC
OCEAN

Murray Springs,
Naco, Lehner Ranch,
Leikem, Escapule

Waco

Trinity

Aucilla River

Friesenhahn

G U L F   O F
M E X I C O

MEXICO    El Cedral

Tepexpan
Mexico City    Tocuila

## MAMMOTHS

Ancestral Columbian

Woolly—skeleton

Woolly—frozen

Columbian

Dwarf

## ARCHEOLOGICAL SITES

Open-air depictions

Mammoth bone-tools

# Guide to Sites

These are some of the most important sites containing mammoth finds ranging from skeletons to cave paintings. Many are no longer accessible; others are only open to research parties; but a few, marked with an asterisk, can be visited by the public, although access may be partial.

**Aa** A small river in northern France where a female woolly mammoth skeleton was excavated in the early 1900s.

**Afar Depression** A region of the Rift Valley in Ethiopia where fossils of early elephants and mammoths, 4–5 million years old, have been recovered from deposits of the Awash Valley.

**Aghnadarragh** A quarry site in Northern Ireland where remains of small woolly mammoths, about 80,000 years old, were recovered in the 1970s and 1980s.

**Ahlen** A German site where the complete skeleton of a woolly mammoth was found; it is now mounted in Münster Museum.

**Angus** The locality in Nebraska where a mammoth skeleton was unearthed in the early 1900s.

**Anza-Borrego Desert State Park**\* Area of southern California, U.S.A., where river deposits have yielded a rich Early Pleistocene fauna including the earliest mammoth skeleton from North America.

**Arcy-sur-Cure** A complex of nine caves in northern France occupied in the Middle and Upper Palaeolithic, two of which (Grotte du Cheval and Grande Grotte) contain depictions dominated by mammoths.

**Aucilla River** River-bottom silts in the Florida "panhandle" that have yielded skeletons of Columbian mammoths and, in 1993, a butchered mastodon tusk.

**Avdeevo** An Upper Palaeolithic open-air site in European Russia. Its fossil assemblage is dominated by mammoths, and the artifacts include portable depictions of the animal, as well as numerous tools made of mammoth bone, and many tools and figurines in ivory.

**Aveley** A site in Essex, near London, England, where the remains of a mammoth were found above those of a straight-tusked elephant in 1964.

**Azov** The town near the Black Sea in southern Russia where one of the world's largest proboscidean fossils, the skeleton of a steppe mammoth nearly 15 ft (4.5 m) high, has been excavated and mounted. A second skeleton was discovered in 1999.

**Baga Oigor** — *see* Tsagaan Salaa

**Baume Latrone, La** A cave near Nîmes, southern France, containing Ice Age paintings done in clay, including seven mammoths, some of which are highly stylised.

**Bechan Cave and Cowboy Cave** Two caves in Utah, U.S.A., in which quantities of Columbian mammoth dung accumulated around 16,000–13,500 years ago.

**Berelekh** The northernmost Palaeolithic site in the world, in northeastern Siberia. It is best known for its "cemetery" of thousands of naturally accumulated mammoth bones, as well as for a unique engraving of an elongated mammoth on a tusk.

**Beresovka** The river in northeastern Siberia near which a well-preserved frozen mammoth carcass was excavated

in 1901, and moved in pieces to St. Petersburg, where its skeleton was mounted and its hide stuffed.

**Bergharen** A former gravel pit in the Netherlands where divers found a mammoth skull 40 ft (12 m) under water in 1994.

**\*Bernifal** A cave in the Dordogne, France, containing a number of Ice Age mammoth depictions as well as "tectiform" signs.

**Big Bone Lick** A site on the Ohio River, Kentucky, U.S.A., whose salty bog soil yielded numerous mastodon bones in the 18th century, and which has some of the latest Columbian mammoth remains in America.

**Bize, Petite Grotte de** A cave in the Aude, southern France, containing five Paleolithic occupation layers, one of which contained a pebble bearing the engraving of a mammoth.

**Blackwater Draw** A series of sites in eastern New Mexico, U.S.A., where the remains of mammoths, camels, horses and bison have been found, plus extensive Paleo-Indian remains, including Clovis points.

**Blanchard, Abri** A rock shelter in the Dordogne, southern France, occupied in the early Upper Palaeolithic, best known for its abundance of mammoth-ivory work, especially its numerous beads.

**Bolshoi Khomus-Yuryakh** River close to Shandrin mammoth site that has yielded a very well preserved woolly mammoth skull.

**Bolshoi Lyakhowski Island** — *see* Liakhov Islands

**Bossilkovtsi** Site in the Russe Region of northeast Bulgaria where remains referred to *M. rumanus* have been recovered.

**Brassempouy** A group of small caves in the Landes, southwestern France, occupied in the Upper Palaeolithic, and famous for a series of mammoth-ivory carvings, including female figurines and a small human head.

**Brno II** An Ice Age burial in Moravia, Czech Republic, which contained a male human skeleton and various animal bones, including a mammoth shoulder blade and two tusks, and a remarkable "marionette" made of ivory.

**Brown Bank** An area of the North Sea between eastern England and Holland whose bed is rich in bones of fauna of the last Ice Age, including woolly mammoths.

**Bruniquel** A series of rock shelters in southwestern France, occupied toward the end of the Ice Age. One of them yielded an antler spear-thrower carved into the form of a mammoth.

**Bzianka** A site in Poland where a 16,700-year-old woolly mammoth skull with abnormal tusks was unearthed.

**Caldas Las** Cave in Asturias, northern Spain, which yielded a sandstone plaquette dating to around 16,000 years ago, with three superimposed mammoths engraved on it.

**Canecaude** A cave in the Aude, southern France, with a number of Upper Paleolithic occupations, the last of which contained a fine antler spear-thrower carved into the form of a mammoth.

**\*Castillo, El** A major cave site in northern Spain, occupied throughout the Palaeolithic, and containing numerous Ice

Age engravings and paintings, including one outline of a mammoth.

**Cernăteşti and Tuluceşti** Sites in the Dacic basin of Romania where remains of the earliest European mammoths, *Mammuthus rumanus*, have been discovered.

**Chabot** A small cave in the Ardèche, southern France, also called the Grotte des Mammouths because of the Ice Age engraved friezes of mammoths and other animals it contains.

**\*Chauvet** A cave in the Ardèche, southeastern France, discovered in 1994, and containing over 400 animal images, including at least 70 mammoths.

**Chembakchino** Site on the Irtysh river of western Siberia from where a skeleton of a steppe mammoth was described in 2004.

**Chilhac** — *see* Senèze

**Clovis** One of the Blackwater Draw sites in New Mexico, U.S.A., Clovis was originally a spring-fed pond where at least 15 mammoths have been found, 6 of them associated with stone points.

**Colby** A stream channel in Wyoming, U.S.A., dating to 12,900 years ago, where parts of at least seven Columbian mammoths were found stacked into piles and associated with artifacts, including stone points and a granite chopper.

**\*Combarelles, Les** A long, narrow cave in the Dordogne, France, containing hundreds of Palaeolithic engravings, including 37 of mammoths (see below).

**Condover** A commercial quarry in Shropshire, England, where one adult and four juvenile woolly mammoth skeletons were discovered in a kettle-hole in 1986.

**\*Cotte de St. Brelade, La** A Paleolithic cave site on Jersey, Channel Islands, containing thousands of artifacts and the stacked bones of 20 mammoths and 5 rhinos.

**\*Cougnac** A decorated Ice Age cave in the Quercy region of France containing paintings, including giant deer, mammoths and ibexes.

**\*Cromer Forest-bed** A complex geological formation in Norfolk and Suffolk, England, containing a rich fossil assemblage ranging from about 1.7 million to 500,000 years ago. Hundreds of molars of steppe mammoths and ancestral mammoths have been collected from localities including Happisburgh, Mundesley, and West Runton.

**Cussac** A cave in the Dordogne (France), near Les Eyzies, discovered in 2000, containing over 150 large engravings dating to c.25,000 years ago. They include a number of fine mammoths with stylistic similarities to those of Pech-Merle Cave.

**Deep Water Channel** An area of the North Sea between Britain and the Netherlands, from which Early and Middle Pleistocene fossils, including *M. meridionalis* and *M. trogontherii*, have been dredged.

**Dent** A Palaeo-Indian site in Colorado, U.S.A., where Clovis points were found in 1932, and which provided the first unequivocal evidence in America for the association of projectile points with mammoth remains.

**Dmanisi** An important Early Palaeolithic site, about 1.8 million years old, in the Caucasus of southeast Georgia.

Bones from six individual humans have been found, including four skulls.

*Dobranichevka* An open-air site on a tributary of the Dnepr River in central Ukraine. Dating to c.15,000 years ago, the site has four mammoth-bone huts, with associated pits, hearths and concentrations of debris.

**Dolní Věstonice** A major open-air Upper Paleolithic site in Moravia, Czech Republic, whose fossils are dominated by mammoth bones, and which has produced a wealth of tools and carvings in ivory, including a famous human head.

**Domebo** Site in Oklahoma, U.S.A., where a scattered mammoth sketon was recovered, together with some Clovis artifacts.

**Domme** A cave in the Dordogne, France, also known as the Grotte de Saint-Front or Grotte du Mammouth because of the large bas-relief carving of a mammoth on its wall.

**Durfort** A site in southern France from which the skeleton of an ancestral mammoth was recovered in the late 19th century. It is now displayed at the Muséum National d'Histoire Naturelle, Paris.

**Eastern Scheldt** A marine inlet in the Netherlands from which numerous mammalian fossils have been dredged, including the teeth of ancestral mammoths.

**Edersleben** A site near Sangerhausen in eastern Germany where a complete skeleton, identified as that of an ancestral mammoth, was excavated.

**El Arco B** A cave in eastern Cantabria, Spain, containing a small archaic engraving of a mammoth, 9 inches (22.5cm) long.

**Elephant Point, Eschscholtz Bay** The site of the first reported find of a mammoth carcass in Alaska in 1907.

**Eliseevich** An Upper Palaeolithic open-air site on the Russian Plain, with mammoth-bone huts, and also containing a limestone carving of a mammoth, and a female figurine in ivory.

**Escapule** A site in Arizona, U.S.A., where a butchered mammoth skeleton was found associated with stone points.

**Eurogeul** A dredged shipping lane west of Rotterdam Harbor, where submerged river sands contain Pleistocene mammal bones that are accidentally caught in fishing nets.

**Fairbanks Creek** The site of the most famous mammoth find in Alaska, U.S.A., where the face, trunk and foreleg of a mammoth calf were uncovered in 1948.

**Fenske** Deposit of lake clay in southeast Wisconsin, U.S.A., where a femur and humerus of a mammoth were found.

**Figuier, Le** A cave in the Ardèche, southeastern France, with Palaeolithic occupations and a few figures on the walls, including stylized mammoths.

*Font de Gaume* An Ice Age decorated cave in the Dordogne, France, with a noteworthy series of mammoths (estimates vary from 26 to 39), sometimes associated with "tectiform" signs.

**Franklin County** Deposits of wind-blown sand near Campbell, Nebraska, U.S.A., in which a huge skull of a Columbian mammoth, now displayed at the University of Nebraska Museum, was found in 1915.

**Friesenhahn Cave** A cave in Bexar County, Texas, U.S.A., where numerous juvenile remains of Columbian mammoths were found alongside those of scimitar-tooth cats.

**Gagarino** An open-air site on the River Don in Russia. Dated to c.25–30,000 years ago, it was the first site (1927) in European Russia where a former structure was detected, with a central hearth and associated pits. A number of ivory "Venus" figurines were unearthed at Gagarino.

*Gargas* A large cave in the French Pyrenees with considerable Paleolithic occupation, and a great deal of art on its walls, including some engraved mammoths.

**Geissenklösterle** A cave in southwestern Germany containing Paleolithic occupations and important ivory figurines from the very early Upper Paleolithic.

**Girl's Hill** Site in Alaska where a flint tool with probable mammoth blood residue was found.

**Gönnersdorf** An extensive late Upper Paleolithic open-air site in northwestern Germany, best known for its 400 engraved stone plaques, including at least 62 mammoth figures.

**Gontsy (or Ginsy)** An open-air site in the Ukraine, discovered in 1871, and dating to c.14,500 years ago. It comprised a series of mammoth-bone huts.

**Gydansk Pensinsula** Large peninsula of north central Siberia where a number of important mammoth discoveries have been made, including the Yuribei carcass in 1977, and the Mongochen skeleton in 2004.

**Hanhoffen** Gravel pits on the Rhine terraces in Alsace (eastern France) that have yielded remains representing various stages in the evolution of *Mammuthus*.

**Hay Springs** A site in Nebraska, U.S.A., where remains of early Columbian mammoths, sometimes identified as *M. imperator*, have been excavated.

**Hebior** Locality in SE Wisconsin, U.S.A., where an almost complete mammoth skeleton, together with some stone artifacts, was found in a lake clay deposit covered with peat.

**Hedehusene** A site in eastern Denmark that has yielded the greatest number of mammoth remains from that country.

**Heilongjiang, Zhaoyuan, Shan Zhan** A silt bed in Harbin, China, where a woolly mammoth skeleton, the most complete in China, was found. It is dated to around 25,000 years ago.

**Helsinki** In the Herttoniemi and Töölö districts, two of only eleven fragmentary mammoth fossils known from Finland have been recovered. Mostly between 30,000 and 18,000 years old, the fossils had been picked up and transported by ice sheets.

**Hohle Fels** A cave in the Swabian Jura of southwest Germany where several ivory figurines (a bird, a half-human, half-animal figure and a possible horse) were recently discovered, dating to more than 30,000 years ago.

**Hohlenstein-Stadel** A cave in southwest Germany in whose Aurignacian levels was discovered a remarkable 11-in (28-cm) ivory statuette of a human with a feline head, dated to about 35,000 years ago.

*Hot Springs* A sinkhole deposit in South Dakota, U.S.A., discovered in 1974, where the remains of at least 50 Columbian mammoth skeletons have been excavated.

**Huntingdon, Utah** Locality in the Wasatch Mountains of central Utah, U.S.A., where an exceptionally well-preserved skeleton of a Columbian mammoth was recovered from lake deposits in 1988.

**Ignatiev** A decorated cave in the southern Urals, Russia, containing several doubtful depictions of mammoths on its walls. Radiocarbon dates indicating calendar ages of 9–7,000 years ago from pigments have cast doubt on previous attributions of the cave's decoration to the Ice Age, and thus further doubt on the identification of some of its figures as mammoths.

**Ilford** An area of northeast London, England, where numerous mammalian fossils, including many mammoth and straight-tusked elephant remains, have been recovered from 200,000-year-old deposits of the River Thames.

**Indigirka River** A large river basin in northeastern Siberia where the frozen remains of a very young woolly mammoth, at least 30,000 years old, were recovered in 1992.

**Ji'nan** City in Shandong, China, where the world's southernmost find of woolly mammoth (36°35'S) was made.

**Jonesboro** A site in Indiana, U.S.A., where the complete skeleton of a Columbian mammoth, now mounted in the American Museum of Natural History, New York, was discovered in 1903. It was originally named as a separate species, *M. jeffersonii*.

**Jovelle** A cave in the Dordogne, France, containing five archaic, deep engravings of mammoths, as well as other animals.

**Kapova Cave** A major cave in Bashkortostan, in the southern Urals of Russia, with Upper Palaeolithic occupation and about 40 paintings of animals and signs on its walls, including at least seven mammoths.

*Kent's Cavern* A large cave near Torquay, southwestern England. One of the first localities to demonstrate the coexistence of humans and extinct animals, it has yielded many artifacts and fossils, including woolly mammoth.

**Khapry** A locality north of the Sea of Azov, southern Russia, where the Liventsovka quarry has yielded abundant early remains of ancestral mammoths around 2–2.5 million years old, which have been named by some authors *M. gromovi*.

**Khatanga** A frozen mammoth carcass found in 1977 on the left bank of the Bolshaya Rassokha River, Siberia, and dated to more than 45,000 years ago. The Jarkov mammoth was found in the same region.

**Kiik Koba** A Middle Palaeolithic cave site in the Crimea, Ukraine, containing two Neanderthal burials.

**Kirgilyakh** A tributary of the Kolyma River, where baby mammoth Dima was discovered in 1977.

**Kniegrotte** An Upper Palaeolithic cave site in eastern Germany, whose artifacts include some unique ivory carvings.

**Kolyma, Middle** The site on the bank of the Bolshaya Baranikha River, Siberia, where a deformed and wrinkled scrap of mammoth trunk, 11 in (28 cm) long, was found in 1924.

**Kostenki** A major group of Upper Palaeolithic open-air sites on the Don River in European Russia, containing large quantities of mammoth bones, including the remains of mammoth-bone huts and bone-filled pits. The sites have also yielded numerous tools and carvings in ivory.

**Kraków, Spadzista** Street A site in southern Poland where the traces of a mammoth-bone hut have been found, dating to about 25,000 years ago.

**Krems-Wachtberg** An open-air site in eastern Austria, which recently yielded two infant burials covered with a mammoth scapula, date to about 30,000 years ago.

**Kyttyk pensinsula** Small peninsula into the Arctic Ocean of northeast Siberia, where some of the latest mainland dates of woolly mammoth have been obtained.

**Lamb Spring** A Palaeo-Indian site in Colorado, U.S.A., where about 40 mammoths and many other species were found.

**Langebaanweg** Disused phosphate quarries in the western Cape of South Africa that have produced hominid and other faunal remains around 5 million years old, including some of the earliest mammoths, *M. subplanifrons*.

**La Sena** Site in Nebraska (near the Franklin County site) with a Columbian mammoth skeleton that has arguably been modified by pre-Clovis humans.

**Lange-Ferguson site** A Palaeo-Indian site in South Dakota, U.S.A., where an adult and juvenile mammoth were possibly killed and certainly butchered using tools made from a mammoth shoulder blade.

**Laugerie Haute** An important Upper Palaeolithic rock shelter in the Dordogne, France, containing a wealth of material including many art objects.

**Lehner Ranch** A site in Arizona, U.S.A., where 13 Clovis points were found with the remains of 13 young Columbian mammoths.

**Lehringen** A Middle Palaeolithic open-air site in Germany best known for the discovery of a broken yew-wood spear between the ribs of a straight-tusked elephant skeleton.

**Leikem** A site in Arizona, U.S.A., where an individual mammoth skeleton was found associated with stone points.

**Lena** The Adams mammoth's frozen carcass was discovered at the mouth of the Lena River, Siberia, in 1799; it was removed to St. Petersburg in 1806.

**Lespugue** A series of small cave sites in the French Pyrenees containing Upper Palaeolithic occupations; one of them yielded a famous ivory female figurine.

**Liakhov Islands** A small male adult mammoth was recovered from the Liakhov Islands, off the north coast of Siberia, in 1901–3, and its skeleton was subsequently given to the Muséum Nationale d'Histoire Naturelle, Paris.

**Lockarp** Site in Sweden that has yielded fragmentary mammoth remains. A tusk found in 1939 is dated to about 15,500 years ago.

**Lourdes** A town in the French Pyrenees containing a number of Upper Palaeolithic caves, one of which, Les Espélugues, yielded a wealth of art objects, including a fine horse carved in mammoth ivory.

**Lovewell** Reservoir in Jewel County, Kansas, U.S.A., where four mammoth skeletons have been excavated from silt deposits since 1969, with debated evidence of contemporary human presence.

**Lugovskoe** A swamp-site in western Siberia, containing remains of at least 27 mammoths, dating to between 19,000 and 14,000 years ago. In 2002, a mammoth vertebra was found here, pierced by a stone spearhead.

**Lynford** River deposit in Norfolk, England, containing numerous mammoth remains and flint tools of Neanderthal type, dating to 60–70,000 years ago.

**Madeleine, La** A large late Upper Palaeolithic rock shelter in the Dordogne, France, containing an abundance of art objects including an engraving of a mammoth on a piece of ivory.

**Maguanjou** A site in northeast China that produced evidence of the earliest humans in eastern Asia at 1.66 million years ago, and which also yielded the earliest known mammoth molars of *M. trogontherii* type.

**Mal'ta** An Upper Palaeolithic open-air site in south-central Siberia, best known for its numerous art objects, including ivory figurines.

**Marche, La** An Upper Palaeolithic cave site in the Vienne region of central France that yielded over 1,500 engraved stone slabs, including some remarkable depictions of mammoths. Recently, parietal engravings, including a mammoth, were found in the adjacent cave of Réseau Guy Martin.

**Mátra** A site in Hungary where the skeleton of a woolly mammoth was unearthed.

**Maxunuokha River** An area of northern Yakutia which has yielded important material of frozen mammoths, including the Yukagir carcass.

**Mayenne-Sciences** A cave in northwestern France, also known as the Grotte du Mammouth, because the Ice Age drawings on its walls include two mammoths.

**Megalopolis** Basin in the Peloponnese, Greece, that yielded a complete skeleton of the ancestral mammoth *Mammuthus meridionalis*.

**Mezhirich** Upper Palaeolithic open-air site in the Ukraine, best known for its five mammoth-bone huts and painted mammoth skull.

**Mezin** Upper Palaeolithic open-air site in the Ukraine, containing a number of mammoth-bone huts, and a series of painted but battered mammoth bones that some believe may represent a set of musical instruments.

**Miami** A Palaeo-Indian site in Texas, U.S.A., where at least five mammoths were found in association with Clovis artifacts.

**Milovice** An Upper Palaeolithic open-air site near Dolní Věstonice in Moravia, Czech Republic, where the remains of a mammoth-bone hut have been discovered.

**Moab** A town in Utah, U.S.A., close to which exists a petroglyph image of a mammoth or mastodon hammered into a rock face.

**Molodova** A group of Palaeolithic open-air sites in western Ukraine, where there is evidence for Middle Palaeolithic mammoth-bone huts.

**Mongocheyaha** River Site on the Gydansk Peninsula of north-central Siberia where a skeleton of woolly mammoth, including a fleshy forefoot and the contents of the animal's stomach, were discovered in 2004.

**Mont Dol** Coastal locality in western France, where hundreds of juvenile mammoth teeth, dating to around 100,000 years ago, have been found in association with stone tools.

**Montopoli** A river deposit near Florence, Italy, where some of the earliest European remains of ancestral mammoths have been recovered.

**Mosbach** A huge sand quarry on the River Rhine in Germany, where thousands of fossils, including numerous remains of the steppe mammoth, have been excavated.

**Mud Lake** Clay deposit in southeastern Wisconsin, U.S.A., where the right lower front leg bones of a subadult mammoth skeleton were discovered.

**Murray Springs** A Palaeo-Indian open-air site in Arizona, U.S.A., where artifacts, including a mammoth-bone shaft-wrench, are associated with mammoth remains.

**Mylykhchan** A site on the Indigirka river in Siberia, where a young frozen mammoth calf was discovered in 1990.

**Naco** An open-air site in Arizona, U.S.A., where 8 Clovis points were found within an adult mammoth skeleton.

**Nemuro Channel** Waterway off eastern Hokkaido, Japan, from which rare fossils of woolly mammoth have been dredged.

**Nogaisk** A site in southern Russia where the complete skeleton of an ancestral mammoth was excavated. It is now mounted in the Zoological Museum in St. Petersburg.

**Obere Klause** A site in southern Germany best known for its late Ice Age engraving of a mammoth on a piece of ivory.

**Obłazowa** An Upper Palaeolithic cave site in southern Poland that contained the world's oldest boomerang – made from a mammoth tusk.

**Oimyakon** District of Yakutia, Siberia, where the head and front part of a mammoth calf were found in a gold mine in 2004.

**Old Crow** A river basin in the Yukon Territory, Canada, which has yielded fossils from the woolly mammoth and earlier mammoth species.

**Olyer Suite** A complex series of deposits in the Kolyma region of northeastern Siberia in which the earliest remains identifiable as woolly mammoths have been found.

**Osaka Group** Middle Pleistocene deposits of central Japan that have yielded remains of *Mammuthus protomammonteus*.

**Otterstadt** Site in southwest Germany where a mammoth mandible with supernumerary molars was discovered.

**Oulen** A cave in the Ardèche, France, decorated in the Ice Age with engravings and paintings, including several mammoths.

**Padul, El** Site in southern Spain that has yielded the southernmost woolly mammoths in Europe.

**Paviland** An Upper Palaeolithic cave site in southern Wales that contained a human burial, accompanied by objects of mammoth ivory.

**Pavlov** An Upper Palaeolithic open-air site in Moravia, Czech Republic, which contained numerous artifacts in mammoth bone and a tusk bearing complex engraved motifs that some interpret as a map or landscape.

**\*Pech-Merle** A major Ice Age decorated cave in the Quercy region of France containing about 60 animal figures, including some remarkable mammoths drawn in black.

**Pietrafitta** A lignite mine in central Italy, which has yielded a rich Early Pleistocene fauna, including 15 crushed skeletons of the ancestral mammoth, *M. meridionalis*.

**Pindal, El** An Ice Age decorated cave by the sea in Asturias, northern Spain, best known for its mammoth painted in red outline with a "heart" inside (see below). A second, smaller outline mammoth figure was discovered a few years ago.

**Polch** A site near Koblenz, Germany, where a partial skeleton of a very old woolly mammoth was excavated.

**Praz Rodet** A Swiss site where a very late mammoth skeleton, only 14,000 years old, was recovered in the 1970s.

**Předmostí** An open-air site in Moravia, Czech Republic. The Upper Palaeolithic occupations are characterized by huge quantities of mammoth fossils (about 1,000 individuals), and a collective grave was covered by mammoth bones. The site is rich in bone and ivory tools, and in portable art, including a mammoth carved in ivory.

**Pushkari I** An open-air site above the Desna river in the Ukraine, with a small longhouse represented by a line of three hearths. The fauna was dominated by mammoth remains, dated from 25,000 to 22,500 years ago.

**Qagnax** A cave on St. Paul Island in the Pribilof group, off the coast of Alaska. Some of the world's latest mammoths, were discovered there.

**\*Rancho La Brea** Area of Los Angeles, California, U.S.A., where thousands of animals, including Columbian mammoths and mastodons, met their death by being caught in tar seeps.

**Réseau Guy Martin** — see **Marche, La**

**Red Crag** Shelly marine sands in eastern England, dating to between about 3.2 and 2.4 million years ago. These include some of the earliest European mammoths, *M. rumanus*.

**\*Rouffignac** A huge Ice Age decorated cave in the Dordogne, France, containing hundreds of painted and engraved figures on its walls and ceilings, heavily dominated by 159 mammoths.

**St. Césaire** A rock-shelter in western France, where the skeleton of a late Neanderthal, about 36,000 years old, was discovered.

**St. Paul** — see **Qagnax**

**Salzgitter-Lebenstedt** An open-air site in northwest Germany, dating to Mousterian (Neanderthal) times around 50,000 years ago. The most noteworthy find is a spearpoint made of mammoth bone.

**San Giovanni di Sinis** Site in western Sardinia, Italy, where remains of a small mammoth have been found.

**Sanga-Yurakh** A river in Yakutia that was the site of the discovery of a frozen female mammoth carcass that was recovered in 1908.

**Sangerhausen** — see **Edersleben**

**Santa Rosa** One of the California Channel Islands, U.S.A., where the remains of dwarf mammoth *M. exilis* have been found.

**Schaefer** Site of a former lake in southeastern Wisconsin, U.S.A., where the majority of a subadult mammoth skeleton was discovered.

**Scoppito (L'Aquila)** Site in central Italy from which a huge ancestral mammoth skeleton was excavated—now housed at the castle of L'Aquila.

**Senèze** and **Chilhac** Two sites in central France from which skulls of the ancestral mammoth have been recovered, dating to around 1.9 million years ago.

**Sevsk** A sand quarry in Bryansk Province, Russia where a natural "cemetery" of mammoths was found in 1988, the biggest known in Europe.

**Shandrin** A tributary of the Indigirka where a mammoth skeleton was unearthed in 1972 containing the digestive organs and stomach contents.

**Shestkovo** Archaeological site in southwest Siberia where the most complete known foetal skeleton of a mammoth was discovered in 1977.

**Siegsdorf** A site in Bavaria, Germany, where Europe's largest woolly mammoth skeleton, nicknamed Oscar, was discovered in 1975.

**Siniaya Balka** A site on the Taman peninsula near the Black Sea, southern Russia, which has yielded abundant remains of mammoths about a million years old, apparently including both ancestral and steppe mammoths.

**Stanton Harcourt** A gravel pit in Oxfordshire, England, where remains of fauna 200,000 years old, including 120 mammoth tusks, have been recovered in ancient deposits of the River Thames.

**Starunia** A site in western Ukraine where rhinoceros carcasses and a mammoth were naturally pickled in a petrochemical seep and surrounded by a mineral wax.

**Steinheim** A site in central Germany where the large skeleton of a late steppe mammoth was excavated in the early 1900s. It is now exhibited in the Stuttgart museum.

**Sungir** An Upper Palaeolithic open-air site in European Russia. The three human skeletons were accompanied by numerous objects, including ivory spears, bracelets, figurines, and about 10,000 beads.

**Süssenborn** Riverine deposits in eastern Germany from which many steppe mammoth teeth have been recovered, including the first to be named as *M. trogontherii*.

**Taimyr** Very large peninsula in north-central Siberia where many important woolly mammoth discoveries have been made, including the 1948 find of an exceptionally complete skeleton, and the celebrated Jarkov mammoth in 1997.

**Tata** A Middle Palaeolithic open-air site in Hungary whose fossil assemblage is dominated by mammoths, and which is best known for a carved and polished segment of mammoth molar.

**Tegelen** A lake deposit in the Netherlands containing a rich fossil assemblage about 1.7 million years old, including the ancestral mammoth *M. meridionalis*.

**Teguldet** A site near Tomsk, western Siberia, where 24 mammoths, including six skeletons, were found during sand quarrying, and have been dated to around 22,500 years ago. Also known as Krasnoiarskai Kuri'a.

**Tepexpan** A site in Mexico that has provided one of the most southerly indications of Columbian mammoths in North America.

**Tocuila** Site near to the Nevada de Toluca volcano in the basin of Mexico, where at least seven Columbian mammoths were found embedded in a volcanic mudflow.

**Tomsk** A site in western Siberia where the broken bones of a young mammoth were found with a fireplace and many stone tools.

**Trinity River** Alluvial deposits near Dallas, Texas, U.S.A., where numerous Columbian mammoth remains have been recovered from sand and gravel pits since the 1920s.

**Trois Frères, Les** A major Ice Age decorated cave in the French Pyrenees, containing hundreds of wall engravings, including a particularly fine mammoth in the "Sanctuary."

**Tsagaan Salaa and Baga Oigor** Two valleys in the Altay Mountains, northwest Mongolia, containing hundreds of petroglyphs, a few of which seem to depict mammoths, and presumably therefore date to the late Pleistocene.

**Tuluceşti** — see **Cernăteşti**

**'Ubeidiyah** A lower Palaeolithic site in Israel including remains of the ancestral mammoth over a million years old.

**Upper Valdarno** An area of the Arno Valley near Florence, Italy, which has yielded abundant fossils of ancestral mammoths, including those that were first named as *"Elephas" meridionalis* in 1825.

**Vilui** A river site west of the Lena River, where a single molar represents the easternmost finding of the ancestral mammoth in Siberia.

**Vogelherd** A Palaeolithic cave site in southwestern Germany where the very early Upper Palaeolithic layers contained carved ivory animal figurines, including some of mammoths.

**Voigtstedt** A site in the Thüringian region of Germany from which some of the latest remains of ancestral mammoths, around 700,000 years old, have been recovered.

**\*Waco** Site in central Texas where 22 Columbian mammoth skeletons were discovered, apparently the victims of periodic flash floods.

**Wally's Beach.** Rich archaeological and palaeontological area of southern Alberta, Canada, where mammoth trackways were discovered.

**Wellsch Valley** A site in Saskatchewan, Canada, containing some of the earliest mammoth remains in North America, about 1.5 million years old.

**West Runton** River deposits in Norfolk, England, which have yielded a skeleton of an early steppe mammoth. Part of the Cromer Forest-bed Formation.

**Wrangel Island** An island in the Arctic Ocean off northeastern Siberia where remains of small woolly mammoths only 10,000–4,000 years old have been discovered.

**Yamal Peninsula** The site of the discovery in 1988 of the baby mammoth Mascha, the westernmost Siberian frozen carcass.

**Yudinovo** Two open sites on the Sudost', a tributary of the Desna, in European Russia. Their age is in the range 19,000–14,000 years ago, and numerous mammoth remains, including three mammoth-bone huts, have been discovered.

**Yuribei** A river in the Gydanskij Peninsula, Siberia, along which, in 1977, a small adult female mammoth was discovered, the latest frozen specimen found so far, dating to 11,500 years ago.

**Yushe Basin** Area of northeast China where remains of the primitive mammoths (*Mammuthus rumanus* and *M. meridionalis*) have been discovered.

**Zemst** A fossil site north of Brussels, Belgium, from which many woolly mammoth teeth, and remains of other Late Pleistocene mammals, have been recovered.

**Zhalainuoer** A coal mine in Inner Mongolia, China, which has yielded China's largest mammoth skeleton, as well as 28½ lb (13 kg) of mammoth dung dating to around 35,000 years ago.

# Bibliography

## General

Agenbroad, L.D. & Symington, R.L. (eds.), 2005. *2nd World of Elephants Congress. Short Papers and Abstracts.* Hot Springs, S.D.: Mammoth Site.

Agenbroad, L.D. & Barton, B.R., 1991. *North American Mammoths: An Annotated Bibliography.* Hot Springs, S.D.: Mammoth Site.

Agenbroad, L.D., Mead, J.I. & Nelson, L.W. (eds.), 1990. *Megafauna and Man: Discovery of America's Heartland* Hot Springs, S.D.: Mammoth Site

Cavarretta, G., Gioia, P., Mussi, M. & Palombo, M-R. (eds.), 2001. *The World of Elephants.* Proc. 1st Int. Congress. Rome: Consiglio Nazionale delle Ricerche.

Cohen, C., 2002. *The Fate of the Mammoth: Fossils, Myth and History.* University of Chicago Press.

Djindjian, F. & Iakovleva, L. (eds.), 2004. *Les Mammouths. Dossiers d'Archeologie* 291.

Foucault, A. (ed.), 2004. *La Vie au Temps des Mammouths. Dossier Pour la Science* hors-série, avril/juin.

Foucault, A., 2005. *Des Mammouths & des Hommes. Deux Espèces Face aux Variations du Climat.* Vuibert: Paris.

Foucault, A. & Patou-Mathis, M. (eds.), 2004. *Au Temps des Mammouths.* Editions du Muséum: Paris.

Garutt, W.E., 1964. *Das Mammut Mammuthus primigenius (Blumenbach).* Stuttgart: Franckh'sche Verlagshandlung.

Guthrie, R.D., 1990. *Frozen Fauna of the Mammoth Steppe. The Story of Blue Babe* University of Chicago Press.

Haynes, G., 1991. *Mammoths, Mastodonts, and Elephants: Biology, Behavior, and the Fossil Record.* Cambridge University Press.

Haynes, G., Klimowicz, J. & Reumer, J.W.F. (eds.) 1999. *Mammoths and the Mammoth Fauna: Studies of an Extinct Ecosystem.* Natural History Museum: Rotterdam.

Joger, U. & Kamcke, C., 2005. *Mammut. Elefanten der Eiszeit.* Braunschweig: Staatliches Naturhistoriches Museum.

Latreille, F. & Buigues, B., 2000. *Mammouth.* Laffont: Paris.

Mol, D. & van Essen, H., 1992. *De Mammoet. Sporen uit de Ijstijd.* The Hague: BZZTôH.

Palombo, M.R., Mussi, M., Gioia, P. & Cavarretta, G. (eds.), 2005. Studying Proboscideans: Knowledge, Problems and Perspectives. *Quaternary International:* pp. 126-128.

Reumer, J.W.F., de Vos, J. & Mol, D. (eds.), 2003. *Advances in Mammoth Research. Deinsea* 9, Rotterdam.

Stahnke, A. (ed.), 2006. *Mensch, Mammut, Eiszeit. Spektrum der Wissenschaft* Spezial.

Storer, J. (ed.) *3rd International Mammoth Conference,* 2003. Program & Abstracts. *Occasional Papers in Earth Sciences* 5, Yukon Palaeontology Program. Yukon.

Storer, J.E. (ed.), 2006. Third International Mammoth Conference, Dawson, Yukon, Canada. *Quaternary International* 142-3.

Surmely, F. *Le Mammouth, Géant de la Préhistoire* Editions Solar, Paris, 1993

Suzuki, N., Tikhonov, A., Agenbroad, L.D. & Lazarev, P. 2005. *Mammoth reborn after 18,000 years: A story of how cutting-edge technology challenged the prehistoric giant.* Newton Press (in Japanese).

Vereshchagin, N.K. & Tikhonov, A.N., 1990. *Exterior of the Mammoth.* Yakutsk Republican Commission for the Study of Mammoths. [in Russian]

West, D. (ed.), 2000. *Proceedings of the International Conference on Mammoth Site Studies.* University of Kansas Publications in Anthropology 22.

## 1 Origins

Agenbroad, L.D., 1998. *Pygmy (Dwarf) Mammoths of the Channel Islands of California.* Hot Springs, S.D.: The Mammoth Site.
———— 2005. North American proboscideans. Mammoths: the state of knowledge, 2003. In: Palombo et al., pp. 73-92.

Bajgusheva, V.S. 2001. Elephants from the delta of the paleo-Don river. In: Cavarretta et al., pp. 172-5.

Cooper, A. 2006. The year of the mammoth. *PloS Biology* 4: pp. 311-313.

Frenzel, B., M. Pécsi and A.A.Velichko (eds.), 1992. *Atlas of Paleoclimates and Paleoenvironments of the Northern Hemisphere. Late Pleistocene – Holocene.* Stuttgart: Gustav Fischer.

Kalb, J.E. and A. Mebrate. Fossil elephantoids from the hominid-bearing Awash Group, Middle Awash Valley, Afar Depression, Ethiopia, *Transactions of the American Philosophical Society* 83, 1993: pp. 1–114.

Kosintsev, P. *et al,* 2004. *Trogontherian elephant from the Lower Irtysh.* Ekaterinburg.

Krause, J. et al. 2006. Multiplex amplification of the mammoth mitochondrial genome and the evolution of Elephantidae. *Nature* 439: pp. 724-727.

Lister, A.M., Sher, A.V., Essen, H. v. & Wei, G., 2005. The pattern and process of mammoth evolution in Eurasia. In: Palombo et al., pp. 49-64.

Maglio, V.J., 1973. Origin and evolution of the Elephantidae. *Trans. Am. Phil. Soc.* 63: pp. 1–149.

McDaniel, G.E. & Jefferson, G.T., 2003. *Mammuthus meridionalis* (Proboscidea: Elephantidae) from the Borrego Badlands of Anza-Borrego Desert State Park®, California: phylogenetic and biochronologic implications. In Reumer et al, pp. 239-252.

Melis, R., Palombo, M.R. & Mussi, M., 2001. *Mammuthus lamarmorae* (Major, 1883) remains in the pre-Tyrrhenian deposits of San Giovanni in Sinis (Western Sardinia, Italy). In: Cavarretta et al (eds.) pp. 481-5.

Osborn, H.F., 1942. *Proboscidea, vol. 2.* American Museum of Natural History, New York.

Palombo, M.R. & Ferretti, M.P., 2005. Elephant fossil record from Italy: knowledge, problems, and perspectives. In: Palombo et al., pp. 107-136.

Poinar, H. et al., 2006. Metagenomics to paleogenomics: large-scale sequencing of mammoth DNA. *Science* 311: pp. 392 –394

Römpler, H. et al., 2006. Nuclear gene indicates coat-colour polymorphism in mammals. *Science* 313: p. 62.

Shoshani, J. and P. Tassy (eds.), 1996. *The Proboscidea: Evolution and Palaeoecology of Elephants and their relatives.* Oxford University Press.

Stuart, A.J. & Lister, A.M. (eds.) (in press) *The West Runton Elephant and its Cromerian environment. Quaternary International.*

Takahashi, T. & Namatsu, K. 2000. Origin of the Japanese Proboscidea in the Plio-Pleistocene. *Earth Science* 54: pp. 257-267.

Takahashi, T. *et al.,* 2006. The chronological record of woolly mammoth in Japan. *Palaeo³* 233: pp. 1-10.

Tassy, P. & Debruyne, R. 2004. Cloner le mammouth? In: Foucault, A. (ed), pp. 38-41.

Vartanyan, S.L., Garutt, V.E. & Sher, A.V., 1993. Holocene dwarf mammoths from Wrangel Island in the Siberian Arctic, *Nature* 362: pp. 337–40.

Wei, G., Taruno, H., Kawamura, Y. & Jin, C. 2006. Pliocene and early Pleistocene primitive mammoths of northern China: Their revised taxonomy, biostratigraphy and evolution. *J. Geosciences,* Osaka 49: pp. 59-101.

## 2 Mammoths Unearthed

Agenbroad, L.D. & Mead, J.I. (eds.), 1994. *The Hot Springs Mammoth Site: A Decade of Field and Laboratory Research in Paleontology, Geology and Paleoecology.* Hot Springs, S.D.: Mammoth Site.

Bongino, J., Nordt, L., Benedict, A. & Forman, S., 2005. Pleistocene stratigraphy and dating of the Waco mammoth site: a basis for environmental reconstruction and interpreting the cause of death. In: Agenbroad & Symington (eds) pp. 27-28.

Buckingham, C. (in press) The context of mammoth bones from the Middle Pleistocene site of Stanton Harcourt, Oxfordshire, England. *Quaternary International.*

Digby, B., 1926. *The Mammoth and Mammoth-Hunting in North-East Siberia.* London: Witherby.

Harris, J.M. (ed.), 2001. *Rancho La Brea: Death Trap and Treasure Trove. Terra* 38/2. Nat. Hist. Mus. of LA County.

Herz, O.F. 1904. Frozen mammoth in Siberia. *Ann. Rep. Smithsonian Inst. for year ending June 30 1903:* pp. 611–25.

Kahlke, R.D. & Mol, D. 2005. *Eiszeitliche Großsäugetiere der Sibirischen Arktis.* Senckenberg-Buch 77. Stuttgart: Schweizerbartsche Verlagsbuchhandlung.

Leroi-Gourhan, A. 1935. Le mammouth dans la zoologie mythique des Esquimaux. *La Terre et la Vie* 2e semestre, 1: 3–12

Lister, A.M., 1993. The Condover mammoth site: excavation and research 1986–93. *Cranium* 10: pp. 61–67

Mashchenko, E.N. et al., 2006. The Sevsk woolly mammoth (*Mammuthus primigenius*) site in Russia: taphonomic, biological and behavioral interpretations. In: Storer (ed.), pp. 147-165.

Mol, D. et al. 2006. Results of the Cerpolex/Mammuthus expeditions on the Taimyr Peninsula, Arctic Siberia, Russian Federation. In: Storer (ed.), pp. 186-202.

Pfizenmayer, E.W. 1939. *Siberian Man and Mammoth*. London: Blackie.

Tolmachoff, I.P., 1929. The carcasses of the mammoth and rhinoceros found in the frozen ground of Siberia. *Trans. Am. Phil. Soc.* 23: pp. 1–76.

Verereshchagin, N.K. 1977. The Berelekh 'cemetery' of mammoths. *Proc. Zool. Inst., Leningrad* 72: pp. 5–50 [in Russian]

Vereshchagin, N.K. & Mikhelson, V.M. (eds.), 1981. *The Magadan Baby Mammoth*. Leningrad: Nauka. [in Russian]

### 3 The Natural History of Mammoths

Boiko, P.V., Mashchenko, E.N. & Sylerzhitskii, L.D., 2005. A new large Late Pleistocene mammoth locality in Western Siberia. In: Agenbroad & Symington (eds.), pp. 22-26.

Buss, I.O., 1990. *Elephant Life: Fifteen Years of High Population Density*. Ames: Iowa State University Press.

Capozza, M., 2001. Microwear analysis of *Mammuthus meridionalis* (Nesti, 1825) molar from Campo del Conte (Frosinone, Italy). In: Cavarretta et al. (eds.), pp. 529-533.

Clarke, E.A. & Goodship. A.E., (in press). A severely disabled mammoth—the palaeopathological evidence. In: Stuart & Lister (eds)

Essen, H. van, 2004. A supernumerary tooth and an odontoma attributable to *Mammuthus primigenius* (Blumenbach, 1799) (Mammalia, Proboscidea) from The Netherlands, and various related finds. *Acta zool. Cracov.* 47: pp. 81-121.

Ferretti, M., 2003. Functional aspects of the enamel evolution in *Mammuthus* (Proboscidea, Elephantidae). In: Reumer et al., pp. 111-116.

Fisher, D.C., 2001. Season of death, growth rates, and life history of North American mammoths. In West (ed.), pp. 121-135.

Fisher, D.C., Fox, D.L. & Agenbroad, L.D. 2003. Tusk growth rate and season of death of *Mammuthus columbi* from Hot Springs, South Dakota, U.S.A.. In: Reumer et al., pp. 117-133.

Gillette, D.D. & Madsen, D.B., 1993. The Columbian mammoth, *Mammuthus columbi*, from the Wasatch Mountains of central Utah. *J. Paleont.* 67: pp. 669-680.

Hoppe, K. 2004. Late Pleistocene mammoth herd structure, migration patterns, and Clovis hunting strategies inferred from isotopic analyses of multiple death assemblages. *Paleobiology* 30: pp. 129-145.

Iacumin, P., Davanzo, S. & Nikolaev, V., 2005. Short-term climatic changes recorded by mammoth hair in the Arctic environment. *Palaeogeography, Palaeoclimatology, Palaeoecology* 218: pp. 317-324.

McNeil, P., Hills, L. V., Kooyman, B. & Tolman, S. M., 2005. Mammoth tracks indicate a declining Late Pleistocene mammoth population in southwestern Alberta, Canada. *Quat. Sci. Rev.* 24: pp. 1253-59.

Mead, J.I., Agenbroad, L.D., Davis, O.K. &

Martin, P.S., 1986. Dung of *Mammuthus* in the arid southwest, North America. *Quat. Res.* 25: pp. 121–27.

Mol, D. (ed.), 2005. *The Yukagir Mammoth: An Animal of the Cold Steppe*. Moscow: European Editions.

Parkman, E. B., 2002. Mammoth rocks. Where Pleistocene giants got a good rub? *Mammoth Trumpet* 18 (1): pp. 4-7, 20.

Pineau, P., 2004. Ce que disent les poils de mammouths. In: Foucault, A. (ed.), pp. 35-37.

Repin, V.E. et al., 2004. Sebaceous glands of the woolly mammoth, *Mammuthus primigenius* Blum.: histological evidence. *Dokl. Biol. Sci.* 398: pp. 382-384.

Spiegeleire, M.A. de, 1985. Figurations paléolithiques et réalité anatomique du mammouth (*Mammuthus primigenius*): essai d'interprétation. *Bull. Soc. belge Anthrop. Préhist.* 96: pp. 93–116.

Suzuki, N., Tikhonov, A., Vereshchagin, N.K. & Hamada, T. 1992. Extracted heart from the frozen baby mammoth in Siberia. *Scientific Papers of the College of Arts and Sciences, Univ. Tokyo*, 42: pp. 63–78.

Suzuki, N., Tikhonov, A., Agenbroad, L. & Lazarev, P. 2005. *Mammoth Reborn after 18,000 Years*. Newton Press: Japan (in Japanese).

Ukraintseva, V.V., 1993. *Vegetation Cover and Environment of the "Mammoth Epoch" in Siberia*. Hot Springs, SD: Mammoth Site.

Verereshchagin, N.K. & Tikhonov, A.N. 1999. The exterior of the mammoth. *Cranium* 16: pp. 1–44.

### 4 Mammoths and Human Culture

Arroyo-Cabrales, J., Johnson, E. & Morett, L., 2001. Mammoth bone technology at Tocuila in the Basin of Mexico. In: Cavarretta et al. (eds.), pp. 419-23.

Aujoulat, N. et al., 2002. La grotte ornée de Cussac – Le Buisson-de-Cadouin (Dordogne): premières observations. *Bull. Soc. Préhist. Fr.* 99: pp. 129-37.

Baffier, D. & Girard, M. 1998. *Les Cavernes d'Arcy-sur-Cure*. Paris: La Maison des Roches.

Bahn, P.G. & Vertut, J. 1997. *Journey through the Ice Age*. London: Weidenfeld & Nicolson. / Berkeley: University of California Press.

Berdin, M.O., 1970. La répartition des mammouths dans l'art pariétal quaternaire. *Trav. Inst. d'Art Préhist. de Toulouse* 12: pp. 181–367.

Bibikov, S.N., 1981. *The Oldest Musical Complex Made of Mammoth Bones*. Kiev: Naukova Dumka. [in Russian]

Bosinski, G., 1984. The mammoth engravings of the Magdalenian site Gönnersdorf (Rhineland, Germany). In: *La Contribution de la Zoologie et de l'Ethologie à l'Interprétation de l'Art des Peuples Chasseurs Préhistoriques* (eds. H-G. Bandi et al.), pp. 295–322. Fribourg: Editions Universitaires.

Bosinski, G. & Fischer, G., 1980. *Mammut- und Pferdedarstellungen von Gönnersdorf*. Wiesbaden: Steiner.

Dobosi, V. T., 2001. Ex Proboscideis— Proboscidean remains as raw material at four Palaeolithic sites, Hungary. In:

Cavarretta et al. (eds), pp. 429-31.

Einwögerer, T. et al., 2006. Upper palaeolithic infant burials. *Nature* 444: p. 285.

Espinoza, E.O. & Mann, M.J. 1991. *Identification Guide for Ivory and Ivory substitutes*. WWF.

Gaudzinski, S., 1999. Middle Palaeolithic bone tools from the open-air site Salzgitter-Lebenstedt (Germany). *J. Archaeol. Sci.* 26: pp. 125–141.

Gaudzinski, S. et al., 2005. The use of Proboscidean remains in every-day Palaeolithic life. *Quat. Int.* 126–128: pp. 179–194.

Gély, B. & Azéma, M., 2005. *Les Mammouths de la Grotte Chauvet*. Paris: Le Seuil.

Gladkih, M.I., Kornietz, N.L. & Soffer, O., 1984. Mammoth-bone dwellings on the Russian Plain. *Sci. Am.* 251 (5): 136–43

Hahn, J. et al (eds)., 1995. *Le Travail et l'Usage de l'Ivoire au Paléolithique Supérieur*. Rome: Istituto Poligrafico e Zecca dello Stato, Libreria dello Stato.

Hannus, A., 2001. Mammoth cutlery: mammoth flakes for mammoth steaks. In: Cavarretta et al. (eds), p. 466.

Haynes, C.V. & Hemmings, E.T., 1988. Mammoth-bone shaft wrench from Murray Springs, Arizona. *Science* 159: pp. 186–87.

Iakovleva, L. & Djindjian, F. 2005. New data on Mammoth bone settlements of Eastern Europe in the light of the new excavations of the Gontsy site (Ukraine). *Quat. Int.*126-28: pp. 195-207.

Jacobson, E., Kubarev, V. & Tseevendorj, D., 2001. *Répertoire des Pétroglyphes d'Asie Centrale. Fascicule 6: Mongolie du Nord-Ouest. Tsagaan Salaa/Baga Oigor*. Paris: De Boccard.

Johnson, E., 2001. Mammoth bone quarrying on the late Wisconsinan North American grasslands. In: Cavarretta et al. (eds), pp. 439-43.

Jordá Cerdá, F., 1983. El mamut en el arte paleolítico peninsular y la hierogamia de Los Casares. In: *Homenaje al Prof. M. Almagro Basch*, 265–72. Madrid: Min. de Cultura.

Kaldenberg, R. L., 2005. Proboscidean & Equine petroglyphs? *Mammoth Trumpet* 20 (3): pp. 17-19.

Khlopatchev, G. A., 2001. Mammoth tusk processing using the knapping technique in the Upper Palaeolithic of the Central Russian Plain. In: Cavarretta et al. (eds), pp. 444-47.

Klein, R.G., 1973. *Ice Age Hunters of the Ukraine*. University of Chicago Press.

Liubin, V.P., 1991. The images of mammoths in Palaeolithic art. *Sovyetskaya Arkheologya* 1991: 20–42. [in Russian]

Mol, D. & Buigues, B. 2005. Notes on the problem of the mammoth ivory trade and the paleontological heritage. In: Agenbroad & Symington (eds.), pp. 124-126.

Münzel, S., 2001. The production of Upper Palaeolithic mammoth bone artifacts from southwestern Germany. In: Cavarretta et al. (eds), pp. 448-54.

Pales, L. & de St Pereuse, M.T., 1989. *Les Gravures de La Marche: IV, Cervidés, Eléphants et Divers.*, Paris: Ophrys.

Penvern, I., 1997. *La représentation du mammouth dans l'art pariétal magdalénien en Périgord: cas particulier de la grotte de*

*Font-de-Gaume aux Eyzies (Dordogne).* Thesis, Muséum d'Histoire Naturelle, Paris. 2 vols.

Pidoplichko, I. G., 1998. *Upper Palaeolithic Dwellings of Mammoth Bones in the Ukraine.* BAR, Int. series 7712, Oxford.

Plassard, J., 1999. *Rouffignac. Le Sanctuaire des Mammouths.* Paris: Le Seuil.

Scelinskij, V.E. & V.N. Sirokov,V.N. 1999. *Höhlenmalerei im Ural. Kapova und Ignatievka.* Sigmaringen: Thorbecke Verlag.

Semenov, S.A., 1964. *Prehistoric Technology* (transl. M.W. Thompson) Bath: Adams & Dart.

Soffer, O., 1985. *The Upper Paleolithic of the Central Russian Plain* Orlando: Academic Press.

Svoboda, J., 2000. Seeing mammoths and using mammoths: evidence from Upper Paleolithic Moravia. In: West, D., (ed.), pp. 153-61.

Tromnau, G., 1983. Ein Mammutknochen-Faustkeil aus Rhede, Kreis Borken (Westfalen). *Archäologisches Korrespondenzblatt* 13: pp. 287-89.

Valde-Nowak, P., Nadachowski, A. & Wolsan, M., 1987. Upper Palaeolithic boomerang made of a mammoth tusk in South Poland. *Nature* 329: pp. 436–38.

### 5 Extinction

Alroy, J., 2001. A mulitspecies overkill simulation of the end-Pleistocene megafaunal mass extinction. *Science* 292: pp. 1893–1896.

Billiard, G., 1947. Y a-t-il encore des mammouths vivants? *Bull. Soc. Préhist. Fr.* 44: pp. 41–43

Brook, B.W. & Bowman, D., 2004. Explaining the Pleistocene megafaunal extinctions: models, chronologies and assumptions. *PNAS* 99: pp. 14624–27.

Brook, B.W. & Bowman, D., 2004. The uncertain blitzkrieg of Pleistocene megafauna. *J. Biogeogr.* 31: pp. 517-523.

Byers, D. A., and Ugan, A. 2005. Should we expect large game specialization in the late Pleistocene? An optimal foraging perspective on early Paleoindian prey choice. *J. Arch. Sci.* 32: pp. 1624-40.

Eiseley, L.C., 1945. Myth and mammoth in archaeology. *Am. Antiq.* 11: pp. 84–87

Fiedel, S. & Haynes, G., 2004. A premature burial: comments on Grayson and Meltzer's "Requiem for overkill." *J. Arch. Sci.* 31: pp. 121-131

Fisher, D.C., 1996. Extinction of proboscideans in North America. In: Shoshani & Tassy (eds.), pp. 296–315.

Fisher, J.W., 1992. Observations on the Late Pleistocene bone assemblage from the Lamb Spring site, Colorado. In: *Ice Age Hunters of the Rockies* (eds. D.J. Stanford and J.S. Day), 51–81. Denver Museum of Natural History.

Fisher, J.W., Jr., 2001. Elephant butchery practices in the Ituri Forest, Democratic Republic of Congo, and their relevance for interpreting human activities at prehistoric proboscidean sites. In: West (ed.), pp. 1-10.

Frison, G.C., 1976. Cultural activity associated with prehistoric mammoth butchering and processing. *Science* 194: pp. 728–30.

——— 1989. Clovis tools and weaponry efficiency in an African elephant context.

*Am. Antiq.* 54: pp. 766–78.

Frison, G.C. & Todd, L.C. 1986. *The Colby Mammoth Site: Taphonomy and Archaeology of a Clovis Kill in Northern Wyoming.* Albuquerque: University of New Mexico Press.

González, S. *et al.*, 2001. Mammoths, volcanism and early humans in the basin of Mexico during the late Pleistocene/Early Holocene. In: Cavarretta *et al.* (eds)., pp. 704-06.

Haynes, C.V., 1966. Elephant-hunting in North America. *Sci. Am.* 214: pp. 104–12.

——— 1991. Geoarchaeological and paleohydrological evidence for a Clovis-age drought in North America and its bearing on extinction. *Quat. Res.* 35: pp. 438–50

Haynes, G., 1989. Late Pleistocene mammoth utilization in northern Eurasia and North America. *Archaeozoologia* 3: pp. 81–108.

Haynes, G. & Eiselt, B.S., 1999. The power of Pleistocene hunter-gatherers. In: Macphee (ed.), pp. 71–93.

Holen, S.R. 2006. Taphonomy of two last glacial maximum mammoth sites in the central Great Plains of North America: a preliminary report on La Sena and Lovewell. In: Storer (ed.) pp. 30-43.

Johnson, C.N., 2002. Determinants of loss of mammal species during the Late Quaternary 'megafauna' extinctions: life history and ecology, but not body size. *Proc. R. Soc. Lond.* B 269: pp. 2221-2227.

Johnson, E. 2006. The taphonomy of mammoth localities in southeastern Wisconsin (U.S.A.). In: Storer, Ed. pp. 58–78.

Levy, S., 2006. Clashing with titans. *Bioscience* 56: pp. 292-298.

Louguet, S. 2005. *Mammuthus primigenius* frm the Middle Palaeolithic site of Mont-Dol (Ille-et-Vilaine, France). In: Agenbroad & Symington (eds.), pp. 89-92.

Loy, T.H. & Dixon, E.J., 1998. Blood residues on fluted points from eastern Beringia. *Am. Antiq.* 63: pp. 21-46.

Lyons, S.K. et al., 2004. Was 'hyperdisease' responsible for the late Pleistocene megafaunal extinction? *Ecology Letters* 7: pp. 859-868.

Macphee, R. (ed.), 1999. *Extinctions in Near Time.* New York: Kluwer/Plenum.

Martin, P.S. & Klein, R.G. (eds.) 1984. *Quaternary Extinctions. A Prehistoric Revolution.* Tucson: University of Arizona Press.

Maschenko, E. N. *et al.*, 2005. New data on the Late Pleistocene mammoth population from the Lugovskoe locality, western Siberia, Russia. In: Agenbroad & Symington, pp.109-113.

Mithen, S., 1993. Simulating mammoth hunting and extinction: implications for the Late Pleistocene of the Central Russian Plain. In: *Hunting and Animal Exploitation in the Later Palaeolithic and Mesolithic of Eurasia* (eds. G.L. Petersen et al.), pp. 163–78.

Owen-Smith, N., 1987. Pleistocene extinctions: the pivotal role of megaherbivores. *Paleobiology* 13: pp. 351–62

Saarnisto, M., 2004. The last mammoths: palaeoenvironment of the Holocene mammoth on Wrangel Island. *Quaternary Perspectives* 14-1: pp. 126-129.

Saunders, J.J., 1980. A model for man-mammoth relationships in Late Pleistocene North America," *Can. J. Anthr.* 1: pp. 87–98.

Schreve, D., 2006. The taphonomy of a Middle Devensian (MIS3) vertebrate assemblage from Lynford, Norfolk, UK, and its implications for Middle Palaeolithic subsistence strategies. *J. Quat. Sci.* 21: pp. 543-556.

Scott, K., 1980. Two hunting episodes of Middle Palaeolithic age at La Cotte de Saint-Brelade, Jersey (Channel Islands). *World Archaeology* 12: pp. 137–52

Soffer, O., 1993. Upper Paleolithic adaptations in central and eastern Europe and man/mammoth interactions. In: *From Kostenki to Clovis, Upper Paleolithic – Paleo-Indian Adaptations* (eds. O. Soffer and N.D. Praslov), pp. 31–49.

Stanford, D., Bonnichsen, R. & Morlan, R.E. 1981. The Ginsberg experiment: modern and prehistoric evidence of a bone flaking technology. *Science* 212: pp. 438–40.

Strong, W.D., 1934. North American traditions suggesting a knowledge of the mammoth. *Am. Anthr.* 36: pp. 81–88.

Stuart, A.J., 2005. The extinction of woolly mammoth (*Mammuthus primigenius*) and straight-tusked elephant (*Palaeoloxodon antiquus*) in Europe. In: Palombo et al., pp. 171–177.

——— et al, 2002. The latest woolly mammoths (*Mammuthus primigenius* Blumenbach) in Europe and Asia: a review of the current evidence. *Quat. Sci. Rev.* 21: pp. 1559-1569.

Waters, M.R. & Stafford, T.W., 2007. Redefining the age of Clovis: implications for the peopling of the Americas. *Science* 315: pp. 1122-6.

Wojtal, P., 2001. The woolly mammoth (*Mammuthus primigenius*) remains from the Upper Palaeolithic site Kraków Spadzista Street (B). In: Cavarretta *et al.* (eds), pp. 367-72.

Wojtal, P. & Sobczyk, K., 2003. Taphonomy of the Gravettian site – Kraków Spadzista Street (B). In: Reumer *et al.* (eds.), pp. 55-62.

Yesner, D.R., Veltre, D.W., Crossen, K.J. & Graham, R.W. 2005. 5,700-year-old mammoth remains from Qagnax Cave, Pribilof Islands, Alaska. In: Agenbroad & Symington, pp. 200-204.

### Interpreting the Evidence

Bocherens, H., 2003. Isotopic biogeochemistry and the paleoecology of the mammoth steppe fauna. In: Reumer et al., pp. 57-73.

Lister, A.M., 1996. Sexual dimorphism in the mammoth pelvis: an aid to gender determination. In: Shoshani and Tassy, pp. 254–259

Lowe, J.J. & Walker, M.J.C., 1997. *Reconstructing Quaternary Environments* (2nd edn.). London: Longman.

Palombo, M.R. et al., 2005. Coupling tooth microwear and stable isotope analyses for palaeodiet reconstruction: the case study of Late Middle Pleistocene *Elephas (Palaeoloxodon) antiquus* teeth from Central Italy (Rome area). In: Palombo et al., pp. 153-170.

# Index

# Acknowledgments

Many friends and colleagues have provided us with invaluable help in the form of documentation, pictures and information. We would particularly like to thank: Larry Agenbroad, Jan Allen, Kathy Anderson, Nikolai Bader, Ian Barnes, Gennady Baryshnikov, René Bleuanus, Odile Boeuf, Gerhard Bosinski, Bernard Bredow, Ian Bricknell, Alan Bryan, Christine Buckingham, Bernard Buigues, Jane Callander, Paul Callow, Lucia Caloi, Chun-Hsiang Chang, Eleanor Clarke, Claudine Cohen, Russell Coope, George Corner, Andrew Currant, Paul Davies, James Dixon, Rob Driscole, Irina Dubrovo, Heidi Eager, Francesco d'Errico, Edgard Espinoza, Hans van Essen, Marco Ferretti, Daniel Fisher, George Frison, Asya Lvovna Gabysheva, Vadim Garutt, Sabine Gaudzinski, David Gillette, Michel Girard, Morris Goodman, Silvia Gonzalez, Wei Guangbiao, Dale Guthrie, Mariana Gvozdover, Erika Hagelberg, Adrien Hannus, Jon Hather, Gary Haynes, Vance Haynes, Victoria Herridge, Gordon Hillman, Steven Holen, Hitomi Hongo, Ludmila Iakovleva, Flavius Ikome, Ann Inscker, Esther Jacobson, Ulrich Joger, Eileen Johnson, Ken Joysey, Kathy Judelson, Ralf Kahlke, Jon Kalb, Alice Kehoe, Wighart von Koenigswald, Thijs van Kolfschoten, Marcel Kornfeld, Pavel Kosintsev, Henryk Kubiak, Nigel Larkin, Francis Latreille, Peter Lazarev, Phyllis Lee, Sergei Lev, Rod Long, Georgi Markov, the late Alexander Marshack, Paul Martin, Evgeny Mashchenko, Geoffrey McCabe, Paul McNeil, Jim Mead, Simon Mikhailov, Elena Miklashevich, Steven Mithen, Dick Mol, Joe Muller, Margherita Mussi, Adam Nadachowski, William Nawrocki, Kate O'Sullivan, Georg Oleschinski, Catherine Orliac, Gilles Pacaud, Shirley-Ann Pager, Maria-Rita Palombo, Isabelle Penvern, Jean Plassard, Philip Powell, Nikolai Praslov, Louise Roth, Dominique Sacchi, Haruo Saegusa, William Sanders, Robert Sattler, Jeffrey Saunders, Kate Scott, the late Nick Shackleton, Viacheslav Shchelinsky, Andrei and Anna Sher, Fyodor Shidlovsky, Vladimir Shirokov, Jeheskel Shoshani, Anthony Stuart, the late Antony and Una Sutcliffe, Richard Tedford, Alexei Tikhonov, Vadim Titov, Pat Troy, Pirkko Ukkonen, Pawel Valde-Nowak, Douglas Veltre, Nikolai Vereshchagin, Yvonne Vertut, Isabelle Walter, Jesse Warner, Jim Whitney, John Wood, Grant Zazula, and Reinhard Ziegler.

We are also grateful to the following colleagues for permission to base graphs on their original work: Christine Buckingam (p.71), W. Dansgaard and colleagues (pp.158–159), Vadim Garutt (pp.174–175), Fiona Grün (p.171), Erika Hagelberg and colleagues (p.42), Steven Mithen (p.161), Kate Scott (p.151), Nick Shackleton and colleagues (p.28) and Tony Stuart (p.148). Paul Bahn visited the cave of Kapova thanks to Olga Boiko, Nick Evans, Viacheslav Kotov, Nigel Lewis and Erika Rauschenbach. Adrian Lister's trip to the Yakutian mammoth exhibition at Dinard was made possible by Gilles Pacaud. We thank Jean Auel for writing the preface; our agents Sheila Watson and Mandy Little; and Antony Mason and the rest of the Marshall Editions team for guiding the book from synopsis to finished product. To anyone whom we may inadvertently have omitted, we extend our apologies.

Illustration Credits
David Ashby 21r, 79t; Bounford Chapman Associates 12t, 16t, 28, 29b, 32, 34, 38b, 53, 65, 92b, 143t, 158–159; Bill Donohue 78–79, 80–81, 96–97, 98–99; Andrew Farmer 114–115, 126–127; Gary Hincks 20–21, 66–67; Mark Iley 142–143, 144–145; Mainline Design 42; Oxford Cartographers 176–180; Richard Phipps 12–13, 14–15, 16–17, 32–33, 36–37; Sue Sharples 97tr; Ann Winterbotham 15b, 67t, 81t, 99t; 147; Michael Woods 5, 13t, 15t, 17t, 22–23, 33t, 37t

Picture credits l = left; r = right; t = top; c = center; b = bottom
Front cover: J. Vertut
Back Cover: Hot Springs, The Mammoth Site of Hot Springs South Dakota
PPages: 1: Francis Latreille/Corbis; 3 M.O. & J. Plassard; 4 A. Sher; 6 P. Bahn; 7 & 8 A. Marshack; 9 A. Sher/Ice Age Museum, Moscow; 10–11 Marshall Editions, London; 22 A. Lister; 23 D. Mol/R. Bleuanus; 24 Marshall Editions; 25 & 26t A. Lister; 26b O. Boeuf; 27tcrl; A. Lister; 27tr A. Sher; 27tcr G. Wei; 27bcr & cr H. van Essen; 27b D. Mol/R. Bleuanus; 28 G. Timonina/V. Titov; 31bl & br A. Sher; 35t A. Sher; 35b Marshall Editions, London; 39 National Parks Service; 40l A. Lister; 40b; 41t & b; 42t Marshall Editions, London; 43 P. Bahn; 44 Mammoth Site of Hot Springs, South Dakota 46 Reuters/Sergei Cherkashin; 47 Photoshot/ItaR-Tass; 48 Getty Images/The Mansell Collection; 49t & b M. Scandella; 50 A. Sher; 51 The Natural History Museum; 52bl & r Torquay Museum; 53t V. Garutt; 53b Novosti; 54 Torquay Museum; 55 Getty Images/Hulton; 56t World Museum of the Mammoth; 56b A. Tikhonov/ZIN; 57 F. Latreille; 58 A. Tikhonov/ZIN; 59 A. Sher; 60bl D. Mol/R.Bleuanus; 60–61t A. Tikhonov/ZIN; 60–6bc G. Zazula; 62t Marshall Editions, London; 62b E. Mashchenko; 63t E. Mashchenko; 63 A. Sher/P. Nikolskiy; 64 & 65t Mammoth Site of Hot Springs, South Dakota; 65b A. Lister; 68 Page Museum; 69t A. Sher; 69b K. O'Sullivan; 70t A. Sutcliffe/Natural History Museum; 70 c & b Flip Schulke; 71 Telegraph Colour Library; 72 N. Larkin; 73t P. McNeil/ S. Tolman; 73b S. Gonzalez; 74t A. Lister; 74b D. Mol/R.Bleuanus; 75l A. Stuart; 75r Norfolk Museums Service/D.Edwards; 76–77; Jean Vertut; 82t & c A. Tikhonov/ZIN; 82b Francis Latreille/Corbis; 83t Francis Latreille/Corbis; 83b M. Newton/Marshall Editions, London; 84 I.Walter/L'Oreal; 85t Masahiro Iijima/Ardea; 85b A. Lister; 86 Torquay Museum; 87tl A. Tikhonov/ZIN; 87tr G. Haynes; 87b M. Barnhart; 88 Museum Autun/World Museum of the Mammoth; 89 A. Tikhonov/ZIN; 90l L. Agenbroad; 90r A. Stuart; 91t M.O. & J. Plassard; 91br P. Bahn; 91bl M.R. Palombo; 92 A. Lister; 93t M. Newton/Marshall Editions, London; 93cl A. Lister; 93b G. Oleschinski; 93r M. Ferretti; 94bl A. Lister; 94r Page Museum; 95t Museum Autun/World Museum of the Mammoth; 95b R. Jacobi; 100 M.O. & J. Plassard; 101t Ardea/Jean Michel Labat; 101b Jean Vertut; 102t E. Mashchenko/Khanty-Mansisk Museum; 102–3b Corbis; 103t P. McNeil; 104l Rev. J. Wood; 104r C. V. Haynes; 105t John Bracegirdle; 105b A. Stuart; 106–107tl D. Fisher; 107tc & 107tr R-D Kahlke; 108t D. Gillette; 108b A. Lister; 109t H. Lumpe/SMNS; 109bl M. Newton/Marshall Editions, London; 110r Hammersmith Hospital, London; 111t A. Stuart; 111b Oxford University Museum; 112–113 French Ministry of Culture; 115 Réunion des Musées Nationaux, Paris; 116 J. Vertut; 117t P. Bahn; 117b Franz Steiner Verlag; 118l Marshall Editions, London; 118br P. Bahn; 119tl D. Sacchi; 119tr Franz Steiner Verlag; 119c & b D. Sacchi; 120t & b Michel Girard, collection La Varende; 121t Corbis/Sygma/CNP/Ministry of Culture; 121b Agence Eurelios/Ministry of Culture; 122 Michel Girard, collection La Varende; 123t P. Bahn; 123c E. Mead; 123b Esther Jacobson/Gary Tepfer; 124t Michel Girard, collection La Varende; 124b P. Bahn; 125t & b J. Vertut; 127 P. Bahn; 128 D. Sacchi; 129t L. Iakovleva; 129b D. Sacchi; 130 P. Bahn; 131 University of Tübingen; 132t A. Hannus; 132bl & br Marshall Editions, London; 133 A. Marshack;134t Marshall Editions, London; 134b N. Bader; 135l & r A. Marshack; 136t, tl, b A. Marshack; 136c University of Tübingen 137t S. Lev; 137b A. Marshack; 138tl & tr M. Newton/Marshall Editions London; 138b Yakutsk Art Museum; 139tl M. Newton/Marshall Editions, London; 139tr E. Espinoza; 139b D.Mol/R.Bleuanus; 140–141 J. Vertut; 145t Marshall Editions, London; 149t A. Stuart; 149c Denver Museum of Natural History; 149b Center for the Study of the First Americans; 150t A. Lister; 150c N. Larkin; 150b Marshall Editions, London; 151l courtesy of the Jersey Post Office; 151r K. Scott; 152t S. Gaudzinski; 152bl E. Mashchenko; 152br E. Mashchenko; 153c Museum of Texas Tech University, Lubbock, Texas; 153b Jim Whitney; 154c & b Marshall Editions, London; 155t A. Lister; 155b E. Mashchenko; 156 Photolibrary/Lon E. Lauber; 157 NHPA/Bryan & Cherry Alexander; 160 D. Gillette; 161 Marcel Kornfeld/George C. Frison Institute of Archaeology & Anthropology; 162 D. Veltre; 163 A. Tikhonov/ZIN.

If the publishers have unwittingly infringed copyright in any of the illustrations reproduced, they would pay an appropriate fee on being satisfied of the owner's title.